THE ENVIRONMENT OF THE EARTH

ASTROPHYSICS AND
SPACE SCIENCE LIBRARY

A SERIES OF BOOKS ON THE RECENT DEVELOPMENTS

OF SPACE SCIENCE AND OF GENERAL GEOPHYSICS AND ASTROPHYSICS

PUBLISHED IN CONNECTION WITH THE JOURNAL

SPACE SCIENCE REVIEWS

VOLUME 28

FRANCIS DELOBEAU

THE ENVIRONMENT
OF THE EARTH

D. REIDEL PUBLISHING COMPANY

DORDRECHT-HOLLAND

L'ENVIRONNEMENT DE LA TERRE

First published by Presses Universitaires de France, Paris

Translated from the French by Janet Rountree Lesh

Library of Congress Catalog Card Number 71–170338

ISBN-13: 978-94-010-3125-7   e-ISBN-13: 978-94-010-3123-3
DOI: 10.1007/978-94-010-3123-3

PREFACE

Among the many works devoted to our space environment, this serious and objective book by Mr. Delobeau should occupy a special place. It has become rare for works on such a subject to be written by competent physicists who are not specialists in the use of space vehicles. While performing research on the ionosphere, Mr. Delobeau was directly involved with the terrestrial environment long before it became common to explore it with sounding rockets and satellites. His professional obligations no longer require him to study aeronomy, only his regular collaboration with a great scientific journal inspires him to keep up to date on this subject. He is particularly motivated by a disinterested appreciation of the information which he hopes to share with his readers. It is a sign of the times that the results of space research should no longer be confined to the circle of space technicians. All of the new tools available to the service of science, for example:—particle accelerators, magnetic resonance, electron microscopy, lasers – have entered the general arsenal following a period of adaptation. Like them, the rocket is now a classical instrument, and gives information even to those for whom it holds no interest in itself.

This book was quite up-to-date when the author submitted his manuscript. Despite the efforts of the editor, this will no longer be completely true when it appears in print. There is scarcely any branch of science that evolves more rapidly than space research. Thus we have an excellent opportunity to reflect upon the question of what we should expect from a work intended, as this one, for the educated but non-specialist reader. Above all, the author should emphasize the essentials of the methods used to obtain results – even more than the results themselves – so that the reader will be able to accept and understand the natural developments in these methods when he learns new results from other sources. It is to Mr. Delobeau's credit that he has, in many cases, thus interpreted his task. However, we must admit that this extrapolation of his book will not go very far; new methods appear daily; fortunately, an explanation of the procedures that have been followed is valuable in itself. In the future, a popularized or general-interest book should be planned as the history of a step in human progress.

For the present, let us congratulate Mr. Delobeau for having familiarized his readers – who we hope will be many – with one of the essential fields of the science of our times.

JEAN COULOMB

Member of the Academy of Sciences

Professor at the Faculty of Sciences of the University of Paris

President of the National Center for Space Studies

TABLE OF CONTENTS

CHAPTER I

THE EARTH IN SPACE

1. Introduction

The Earth is not confined to a nearly spherical solid form 6370 km in radius, isolated in space, turning on its axis once in 24 hours and around the Sun once in 365 days. It has a very complex environment which extends to considerable distances, ending only where the solar environment begins.

This environment includes:

– *A gravitational field* which is governed by the law of universal gravitation. Theoretically such a field, which decreases as the inverse square of the distance, has no true limit. However, it can be considered to end at the point where the attraction of other heavenly bodies becomes dominant. Leaving the Moon out of consideration, we can say that this effect sets in at about 8×10^6 km.

– *A magnetic field* which has a spatial configuration similar to that which would be produced by a bar magnet passing almost through the center of the globe and aligned nearly along the axis of the geographic poles. This field varies as the inverse cube of the distance. Therefore it also extends theoretically to infinity. In practice, it is confined to a so-called 'geomagnetic cavity' whose form is not yet precisely known. In the direction of the Sun, its radius is about ten terrestrial radii.

– *A vast gaseous layer* which is given the generic name of atmosphere. Although its mass is very small – only a ten-thousandth of that of the Earth – the atmosphere plays an essential role in determining the conditions that prevail on the Earth's surface, which it protects from radiation incompatible with the presence of life.

The terrestrial atmosphere has a structure that depends essentially on altitude. In the first approximation, its density decreases exponentially, but above a certain altitude its composition changes.

The molecules of which it is composed become dissociated into atoms, which in turn are ionized under the influence of solar radiation. The components, which were always influenced by the gravitational field, then become subject to the magnetic field as well. Scattering phenomena come in. It becomes difficult to speak of a definite limit. One has to distinguish between the neutral gas and the ionized gas, and introduce a dynamical model. Neutral particles are present in significant numbers only up to altitudes of 500 to 600 km, while ions and electrons, whose motion is essentially governed by the Earth's magnetic field, are confined by the solar wind and the interplanetary magnetic field. These last two phenomena, whose existence had long been suspected, were discovered only in the course of the last few years.

The terrestrial atmosphere should be imagined as undergoing a continuous evaporation at its outer surface, which is different for neutral and for charged particles.

The terrestrial environment is subject to outside influences. In addition to the electromagnetic radiation from the Sun, which we have already mentioned, it is continuously influenced by charged particles emitted by the Sun, by meteorites of widely varying sizes that come from the solar system, and by cosmic rays.

In this book, we shall spend relatively little time on the very low atmosphere, between the ground and an altitude of about fifty kilometers, for its characteristics are not fundamentally different from those at ground level – with the exception of an ozone layer to which we shall devote some attention.

2. Historical Survey

Since the most ancient times, the existence of air has been known from its motion. It was long considered to be one of the fundamental components of matter. But certainty as to the sphericity of the Earth is much more recent. Although this idea was asserted by Pythagoras and Aristotle, it found little acceptance thereafter, for want of a definitive proof. The difficulties encountered by Galileo are famous.

The invention of the barometer goes back only to 1643. It led to the now common-place notion of atmospheric pressure. Since the direct effects of this phenomenon are rarely apparent in our surroundings, it is not surprising that it was discovered only comparatively recently. Pascal showed that this pressure is none other than the weight per unit surface area of the air situated above the level at which the measurement is made. He concluded that it should decrease with altitude. And he verified this fact in the famous experiments which he carried out in 1648 at the summit of the Puy de Dôme, and which he repeated in Paris. This result put an end to the thesis that nature abhors a vacuum, and marked the true point of departure for the study of the terrestrial environment. Later on, it was determined that the temperature also decreases in a regular fashion, and it was thought until around 1900 that this decrease continued up to a peak that was arbitrarily placed at about 50 km above the ground.

It was not until 1774 that the work of the chemist Lavoisier made it possible to determine the composition of air. What we breathe is a mixture of two gases in proportions that are practically independent of place and time. Besides these major constituents – molecular oxygen and nitrogen – air also contains the rare gases, carbon dioxide, water vapor, and traces of many other components. Argon, the most abundant of the inert gases, was discovered only in 1892 (Lord Rayleigh).

Table I indicates the distribution by volume of the principal components, at ground level.

Two methods for obtaining data far from the surface of the Earth were available *a priori:*

– making measurements on the spot;

– deducing facts concerning the nature of the upper layers from certain effects which could be studied at ground level.

Until the Second World War, the first approach remained extremely limited. It was, however, frequently used up to modest altitudes; and our knowledge has progressed by means of alternate developments in these two research methods.

For a long time, the only direct method of observation consisted of transporting apparatus to the summit of a mountain. In 1783, the first balloon ascent by the Montgolfier brothers opened new perspectives; and during the nineteenth century, new altitude records were constantly being established. The presence of a passenger placed a severe limit upon the potential of this approach, but in 1935, at the time of the famous expedition of the balloon Explorer II, the techniques perfected by Prof. Piccard made it possible for a man in a gondola to reach the altitude of 22.5 km – a remarkable feat for the time.

Aviation made its appearance at the beginning of the present century. It has made possible a systematic study of the statics and the dynamics of the very low atmosphere

TABLE I

The composition of air at ground level

Component	Volume (%)	Component	Volume (%)
Nitrogen (N_2)	78.08	Neon (Ne)	1.82×10^{-5}
Oxygen (O_2)	20.95	Helium (He)	5.2×10^{-6}
Argon (A)	0.93	Krypton (Kr)	1.1×10^{-6}
Carbon dioxide (CO_2)	0.03	Xenon (Xe)	1.0×10^{-7}

Number of particles per cubic centimeter: 2.9×10^{19}.
Normal atmospheric pressure: 101.625 pascals.

and has stimulated the development of meteorology. However, it could not lead to any great progress as far as the upper atmosphere is concerned.

The sounding balloons which were born almost at the turn of the century offered further possibilities. The measurements they collected *in situ* were recorded on board and recovered after descent, with an increasing probability of success. The discovery of an isothermal zone by means of these balloons threw into disarray the simple concepts which had been assumed up to that time. These balloons are still frequently used because of their modest cost.

With the development of radio techniques, the sounding balloons were transformed into radio sounders.

In principle, they eliminate any loss of information, because the data are immediately re-transmitted to receiving stations. The ceiling for radio sounders is about 20 km. The measurements they make (pressure, temperature, degree of humidity) are necessarily affected by systematic errors which the experimenters had to learn how to estimate, and for which the corrections are now well understood.

At the same time, the geophysicists made an effort to obtain some insight into the higher levels of our atmosphere from the study of physical phenomena at ground level. Let us give two examples:

– Very powerful explosions cease to be heard at distances greater than about 100 km. Amplitude damping is not the cause of this phenomenon. The only possible explanation lies in the influence of the vertical thermal gradient upon the speed of sound. The waves curve inward towards higher altitudes and are reflected from a hotter layer; whence we deduce the existence of a zone of silence beyond which it is possible to hear again.

– The terrestrial magnetic field undergoes normal and anomalous variations with time. Although the principal origin of this field is internal (the circulation of electric charges in the heart of the Earth), we are led to infer the existence of an external contribution due to charged particles in motion around the Earth. It is not possible to specify the altitude and size of the stream by this method, but we can deduce that the components of air are ionized above a certain level.

A much more efficient line of research appeared in 1901. In that year, Marconi received a radio signal which had traveled more than 3000 km. This discovery was a great surprise, for in a neutral, homogenous medium radio waves propagate in a straight line away from their center. It was not long before explanations were offered. One of them, formulated independently by Heaviside and Kennely, proved to be correct. It modified all the previous ideas on the more or less uniform composition of the upper atmosphere, while doing justice to a few precursors like Balfour Stewart, who had predicted these phenomena as early as 1882.

If distant signals can be received in this way, there must exist a region which acts as a concave mirror, reflecting the waves back to the ground. Successive reflections between this mirror and the Earth make it possible, within the limits imposed by damping, to effect radio connections at large distances all around the Earth.

It was out of the question at that time to proceed with systematic studies to determine the properties and the origin of this mirror – in particular, its altitude, its reflecting power, and its temporal or geographical variations. Twenty-five years later, progress in electronics enabled Appleton and Barnett, and then Breit and Tuve, to prove that the upper part of the atmosphere is ionized. As Heaviside himself suggested, this ionization is caused by short-wavelength solar radiation absorbed by the components of a rarefied medium.

The echo technique – which consists of illuminating the mirror radar-fashion by directing towards it radio waves of variable frequency chopped into pulses – has been a powerful research tool for more than thirty years. It has put within our reach a considerable section of the atmosphere, extending from 60 to 300 km. It is to this method that we owe a large part of our detailed knowledge of the lower *ionosphere*, or ionized portion of the atmosphere. The method continues to be widely used from fixed sounding stations, which have been established even at almost inaccessible sites on the globe.

But the true conquest of the upper atmosphere could be accomplished only *in situ*. Now, Archimedes' principle – which is the basis for the operation of balloons and radio sounders – offers only limited possibilities. In order to do much better – that is, to exceed about thirty kilometers with a substantial load – one has to call upon another law of mechanics: that of the conservation of momentum in isolated systems.

If an instrument package is connected to a source of chemical – or even nuclear – energy, which liberates its energy in thermal form, a considerable momentum is imparted to the gases resulting from the reaction. There results an autopropulsion in the opposite direction. The surrounding medium plays no role, except for braking the motion.*

Direct explorations by rocket began at the end of the Second World War, thanks to the V-2 rocket developed by the Germans during the conflict and modified by the Americans in order to obtain maximum altitude rather than maximum range. Vehicles better adapted to geophysical research were produced later. First there was the 'WAC Corporal', a two-stage rocket capable of reaching 400 km, and then the 'Viking'. Other models were launched from balloons, in order to increase the altitude by eliminating much of the air resistance. Later on, other rockets appeared in the United States, in the U.S.S.R. and in other countries, capable of carrying apparatus to altitudes of more than 1000 km.

However, this method of observation is still not the ideal means of studying the high and the very high atmosphere. The altitude changes rapidly and the duration of the flight is only a matter of minutes. The many geographic, magnetic, solar, and other parameters to be observed require a very large number of flights.

In October 1957, the thrust[†] developed by a rocket became great enough to impart to a vehicle of significant mass the minimum velocity for Earth orbit. This concept is a direct result of the law of universal gravitation. If an object describes an orbit (which we shall assume, for simplification, to be circular) around the Earth, its centrifugal force balances its weight.[‡] Therefore there exists at each level an appropriate tangential velocity: 8 km/s at an altitude of 200 km, for example. This is called the first cosmic velocity – as opposed to the second cosmic velocity, which is the velocity necessary to escape from the Earth's attraction.[§] On account of the dense layers of air near the

---

* If $m(t)$ is the mass of the vehicle (which decreases with time), u the velocity of the ejected gases and v the velocity of the rocket, we can write for each time, t,

$$m\frac{dv}{dt} = -u\frac{dm}{dt},$$

whence $v(t) = uL(m_0/m(t))$, if m_0 is the initial mass. The velocity u must be substantial (2.5 km/s). The velocity v can be greater than u. Large velocities are obtained only by consuming a great deal of mass; this is the reason for the two- and three-stage rockets which make it possible to dump in mid-flight the masses of structures which have become useless after the reactants have been consumed.

† The *thrust* is the force which the propellant applies to the object – that is, $u(dm/dt)$ in the preceding notation. The impulse is the product of this quantity and the time during which the thrust is applied. Reduced to unit mass, it becomes the *specific impulse*, which is analogous to a velocity. It is often seen expressed in seconds. This is the result of a confusion between weight and mass, and amounts to introducing the acceleration of gravity at the surface of the Earth into questions where it is quite irrelevant.

‡ Let mg be the weight of the vehicle, where m is its mass and g the acceleration of gravity at a distance r from the center of the Earth. Setting this force equal to the centrifugal force mv^2/r, one obtains the velocity $v = \sqrt{gr}$ of the vehicle in its trajectory.

The mass no longer enters after injection into orbit. However, it plays an important part in determining the capacity of the rocket necessary for launching.

§ The escape velocity can be obtained at once by writing that we must furnish at the outset a kinetic energy $\frac{1}{2}mv^2$ equal to the potential energy of the vehicle at infinity, $\int_0^\infty mg\,dr$. Again, the mass cancels out. We find $v_0 = \sqrt{2g_0a}$, where a is the radius of the Earth and g_0 the acceleration of gravity at the Earth's surface. Inserting the correct values, we find $v_0 = 11.18$ km/s.

ground, there exists in practice a minimum orbital altitude, on the order of 180 km, below which the lifetime of the satellite is very short.

It was, as we know, the Russians who placed the first artificial satellite into Earth orbit. In the last twelve years, enormous progress has been made in this field, and hundreds of space probes have been launched for very diverse purposes. Research programs have been planned and accomplished. They have multiplied without end and have caused our knowledge of the upper atmosphere to grow in a spectacular fashion. All the barriers of the terrestrial environment have been crossed, and we have become aware of the position of the Earth in space. Many false concepts have been abandoned. Some phenomena have been explained, while others have been discovered.

We can say that, on the whole, all the information deduced from studies made at or from the ground has been confirmed. We have obtained access to the upper portion of the ionosphere, which is in practice beyond the reach of radio-sounding techniques. The 'Van Allen belts' were discovered as early as 1958, and a picture was formed of the magnetosphere – the region of space which is the site of the Earth's magnetic field – and of a permanent shock wave due to the action of the solar wind and located around 100 000 km from the center of the Earth.

3. The Structure of the Atmosphere

The description of the terrestrial environment can be approached in different ways. Various classifications can be adopted, depending upon the physical processes invoked. First of all, a review of the numerous relevant parameters is indicated. Let us then consider the gas of the atmosphere, contained in the gravitational and magnetic fields of the Earth, and subjected to both radiative and corpuscular emission from the Sun. All these factors introduce spatial and temporal variations which can be either regular or random. It is therefore necessary to distinguish – especially for the uppermost part of the atmosphere – between a so-called normal state and a temporarily disturbed state. For simplification, we shall consider here only the ordinary static and dynamic condition, leaving the study of the various types of disturbance for a special chapter (cf. Chapter VI).

(a) Gravitation can be considered as a purely radial force. Every particle is attracted with a force proportional to its mass, directed towards the center of the Earth, and rapidly decreasing with distance. We write that

$$\mathbf{p} = m\mathbf{g} \quad \text{with} \quad \mathbf{g} = -\frac{f\mathrm{M}}{(a + h)^2}\mathbf{i},$$

where f is the constant of universal gravitation, M the mass of the Earth, a its radius, and h the altitude. At ground level, $h=0$ and $g=g_0=9.8$ m/s^2. The decrease in g can often be neglected; indeed, one has to reach an altitude of 2500 km before the gravitational force is divided by two.

Placed in such a field, particles distribute themselves according to a well-known law which fixes the decrease in density with increasing altitude. This decrease depends upon the nature and the temperature of the components. But air is a mixture. As long as the density is high enough, this mixture remains homogeneous, on account of the incessant collisions and the fact that the molecular weights of nitrogen and oxygen are similar.

This is no longer the case at higher altitudes. The much lighter helium and hydrogen atoms make their appearance above a certain level, and become the principal components. This can be understood from the fact that in collisions between heavy and light particles, upward scattering favors the latter.

The law of gravitational attraction is apparent in another phenomenon: that of the tides, which are to be found in the atmosphere as well as in the oceans. At ground level, the change in atmospheric pressure with time is evidence for this periodic phenomenon, which is even more pronounced at higher altitudes. The Sun and the Moon play an important part in determining this periodicity.

(b) The Sun is the principal source of energy for the atmosphere, at all altitudes. It imposes cyclic variations of two kinds:

– those connected with its own intrinsic condition. They are marked by a periodicity of twenty-seven days, corresponding to the Sun's rotation on its axis, and by a much longer periodicity of eleven years, which we shall consider later.

– those connected with its position with respect to the terrestrial observer. They appear as daily and seasonal effects, and depend upon the geographic latitude.

The Sun delivers its energy in two forms: radiative and corpuscular. The first is most important in the interior of the atmosphere. The second concerns principally the outer reaches, at least when the Sun is quiet.

Solar electromagnetic radiation is familiar to us from visual observation. Its emission spectrum is, however, much more extended than the region that can be detected at the ground by the eye or any other sensitive instrument. It extends from centimeter wavelengths to X-rays, which are measured in ångström units ($1\text{Å} = 10^{-10}\text{m}$).

The corresponding photons, poured out in enormous quantities,* radically alter the composition of the atmospheric medium. They act in a selective fashion on the components of air, each atom being characterized by well-determined excitation and ionization energies. In particular, they produce:

– the dissociation of molecules into atoms;
– the excitation and ionization of these atoms.

Thus there are created new neutral particles, positive ions and electrons. As the electrons can also attach themselves to neutral particles, negative ions are observed as well. These particles undergo a large number of interactions with the other atoms, molecules, ions and electrons, and photons, producing exchange of charge between the participants, or new excitations and ionizations, or simple elastic collisions, or even ion–electron recombinations.

Not all the incident photons give rise to such events. They must have a well-determined energy to act upon a particle of a given kind. Fortunately, those whose

---

* At the top of the terrestrial atmosphere the photon flux is on the order of 10^{14} photons/cm²/s for wavelengths shorter than 2000 Å.

associated wavelength* is in the visible region penetrate to the ground. It is the ultra-violet and X-ray photons that are captured, and these captures occur at different altitudes.

Schematically, we can distinguish three components:

– the neutral particles, untouched by solar radiation or produced by the recombination of charged particles;

– the heavy, positively charged ions;

– the light, negatively charged electrons.

In addition to their random motions, these particles can have a stream motion produced by temperature gradients and by the terrestrial magnetic field.

(c) The existence of the Earth's magnetic field has been known for a long time. It is very weak compared to those which are created every day in the laboratory, but it nevertheless has a very important effect. Its role becomes dominant at very high altitudes. The neutral particles are completely unaffected by it, but the ions and electrons are subjected to forces which tend to make them follow complicated trajectories, which become longer with decreasing frequency of collisions. The distribution of this magnetic field introduces new parameters such as the intensity of the magnetic field, the latitude and the magnetic inclination which, as we shall see, determine the behavior of the atmosphere above an altitude of 250 km.

Of all the parameters we have reviewed, only the altitude influences all the different phenomena. This will therefore be our criterion for a general description of the atmosphere.

4. The Neutral Gas

A. COMPOSITION

The neutral atmosphere is characterized by its composition, mean molecular weight, density and temperature.

At its base, as we have seen, are found almost exclusively the nitrogen molecule N_2 and the oxygen molecule O_2. Water vapor obviously plays an important part; but this is a 'surface' phenomenon, negligible on the scale we are adopting.

The composition of air remains almost unchanged up to an altitude of about 100 km. The components have similar molecular weights, and remain well mixed. This situation changes with the dissociation of the O_2 molecule, which becomes appreciable above about 100 km (see Figure 1). Molecular nitrogen is much more stable and can still be found in appreciable quantities above 1000 km, although atomic oxygen rapidly becomes the major constituent. Helium, whose concentration decreases only slowly above a few hundred kilometers, finally becomes dominant. It is in its turn replaced, little by little, by the even lighter atomic hydrogen.

In addition, there are photo-chemical reactions stimulated by solar radiation, which result in the creation of ozone (O_3) and of other minor components such as NO, OH,

---

* The energy E of a photon is related to the associated wavelength by the formula $E = hc/\lambda$, where h is Planck's constant (6.62×10^{-34} J s) and c is the velocity of light (3×10^8 m/s).

and H_2O. The mean molecular mass therefore decreases with altitude beyond about 100 km.

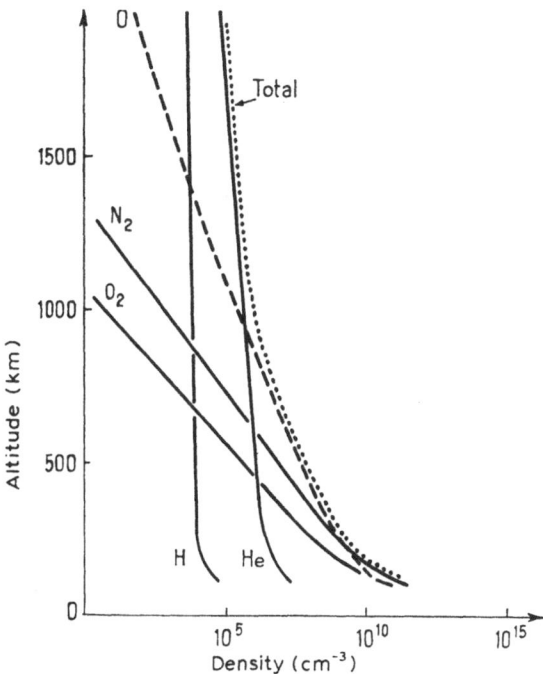

Fig. 1. Distribution of the constituents of the atmosphere as a function of altitude.

B. TEMPERATURE

Temperature is a thermodynamic concept which characterizes the state of equilibrium of a gas. This equilibrium, in the statistical sense, is brought about by the numerous collisions due to thermal agitation. It requires a 'thermalization time' which is very short compared with the characteristic times for temporal change in the atmosphere, at least so long as the medium is sufficiently dense. The temperature is defined up to a critical level h_c which can be estimated to be about 500 km.

Before the appearance of satellites, temperature and composition had to be estimated from spectroscopic observations made from the ground, or from the rare results obtained by rocket. Since then, great progress has been made, especially above 200 km, by the American *Explorer* and *Vanguard* satellites. The terminology introduced by S. Chapman has now been officially adopted. It depends principally upon the vertical distribution of the temperature.

Starting from the ground, we encounter in succession (Figure 2):

– *the troposphere*, which is heated by the Earth's surface and has its ceiling at around 15 km. In this region, the temperature decreases linearly to 200 K;

– *the stratosphere*, which is separated from the preceding layer by the tropopause. In this region the temperature increases until it regains its ground-level value at around 50 km, the level of the stratopause. The energy required for the reheating is obtained from the solar radiation;

– *the mesosphere*, marked by a renewed drop in temperature. It is relatively unstable, and stops at an altitude of 85 km (the mesopause).

These three regions have comparable hydrodynamic characteristics.

Going farther up, we enter the *thermosphere*, the site of a temperature increase – very rapid, at first – and a profound change in the nature of the medium. A fraction of the solar energy absorbed produces the reheating.

Beyond 500 km, the temperature no longer depends on the altitude to any great extent, but it admits of strong diurnal variations. Collisions between particles become so rare that the laws of hydrostatics cease to be applicable.

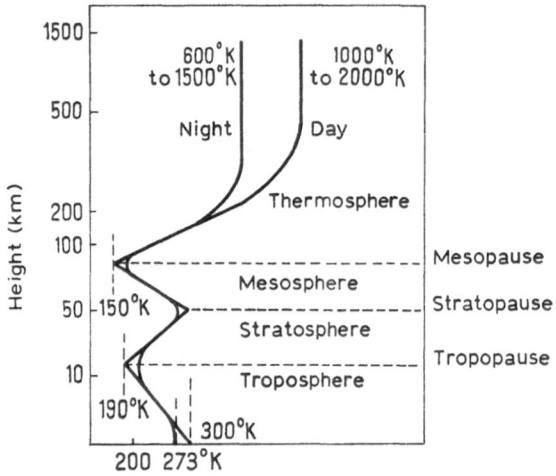

Fig. 2.   Regions of the atmosphere divided according to temperature (after M. Nicolet).

As a result of their last collision, certain particles will have a velocity suitable in magnitude and direction to describe a ballistic trajectory. This is what we call any orbit followed by an isolated body. It is easily demonstrated that, in a gravitational field, these curves are conics – ellipses, hyperbolas, or parabolas. If the velocity is less than the escape velocity at the level under consideration, the particle will describe an ellipse which will bring it back to the lower, denser atmosphere after describing a path whose length is determined by the next collision. There is thus a sort of sustained boiling with partial evaporation.

The *exosphere* is the upper part of the atmosphere, considered from this point of view. It begins at the altitude h_c above which ballistic trajectories begin to dominate. It is therefore impossible to speak of a distinct boundary for the neutral gas.

C. PRESSURE AND DENSITY

In the first approximation, it is easy to find a law governing the decrease of the density or pressure of the particles in the atmosphere as a function of altitude. We have only to write that the weight is balanced by the pressure gradient, that is,

$$\rho g + \frac{dp}{dh} = 0 \, ,$$

where ρ is the density.

The pressure and the density are related by a simple, well-known equation of state: the perfect gas law, which is written

$$p = \rho \frac{kT}{m} \, ,$$

where k is Boltzmann's constant (1.38×10^{-23} J/K), m is the mean molecular mass of the gas, and T is its temperature.

Eliminating ρ, we find

$$\frac{p}{H} + \frac{dp}{dh} = 0 \, ,$$

where we have set

$$H = \frac{kT}{mg} \, .$$

This parameter has the dimensions of length, and is called the scale height. It is very important, for it tells us how fast the pressure decreases. We see that it also represents the thickness that the atmosphere would have if the pressure were everywhere equal to its value at ground level. With the classical values $M = 29g$, $g_0 = 9.8$ m/s^2 and $T = 300$ K, we find $H = 8.5$ km.

If we want more information, we have to make some hypotheses. The simplest is to assume that H is constant. It is then easy to obtain

$$p = p_0 e^{-h/H} \, ,$$

where p_0 is the ordinary atmospheric pressure.

It is interesting to note that we thus obtain a decrease in pressure by a factor of 10^6 at an altitude of 100 km, and that this result is in good agreement with the measurements.

Our difficulties begin beyond this point, for then the scale height can no longer be approximated by a constant. That is why we give the name of *homosphere* to everything located below an altitude of 100 km, reminding ourselves that this is a quasi-homogeneous medium.

The height H varies with temperature, molecular weight and gravity. This latter factor plays only a small role, as we have seen, so long as the altitude represents a small fraction of the Earth's radius. On the other hand, M and T vary a great deal in the region we shall call the *heterosphere*.

No estimate is possible without a few measurements. These have been made by satellite since 1958. They show that the pressure is much more nearly constant than is indicated by the preceding exponential law. We conclude that H increases on account of

– higher temperatures;
– lower molecular weights (dissociation of O_2).

Figure 1 indicates the distribution of the components of the atmosphere, in the light of our present knowledge. Figure 3 summarizes the distribution of pressure, density, temperature, and scale height. Note that this last quantity becomes large above 500 km.

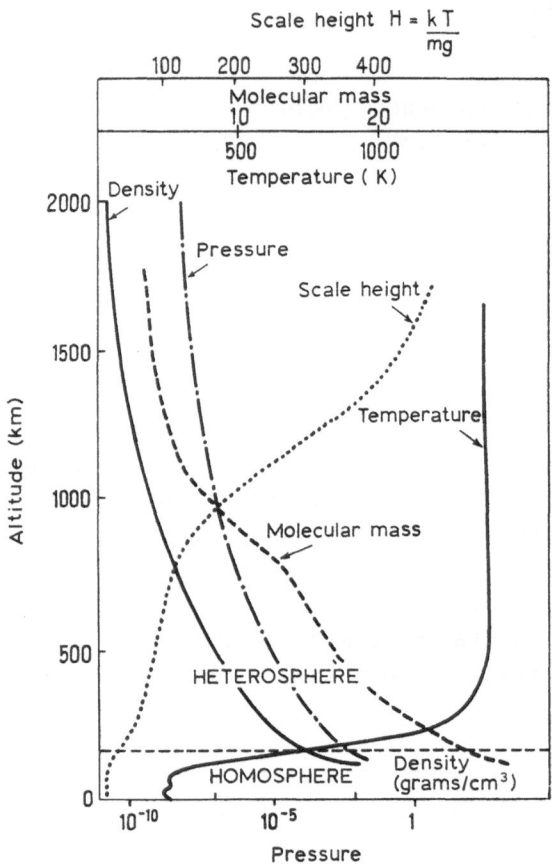

Fig. 3.   Distribution of the parameters of the atmosphere.

5. The Ionized Gas

From 50 km upwards, the atmospheric components begin to undergo photoionization. A fraction of the neutral particles is therefore split into positive ions and electrons. This new level separates two new regions superimposed on the preceding ones: the *neutrosphere* and the previously mentioned *ionosphere*.

Photoionization is a reaction of the type

$$n + h\nu \rightarrow n^+ + e \,,$$

where n designates the neutral atom and $h\nu$ the photon. The ionosphere remains electrically neutral, from the macroscopic point of view. Ions and electrons can recombine by an inverse reaction, as the result of a collision. In general, no photon is re-emitted. The corresponding energy is usually released in kinetic form.

The ionosphere has been divided into *regions*, which may themselves contain several *layers*. This terminology has been adopted in order to show the nature of the mechanism which governs their creation. In principle, each region is the site of the photo-ionization of a different component. In practice, there exist other, more complicated, processes. The existence of a region is easily imagined by considering one of these processes whose maximum ionization occurs at a characteristic altitude. Above this point there is a lack of ionizable constituents and an excess of incident solar photons. Below, the reverse is true. We shall discuss these mechanisms in more detail in Chapter III.

The *D region* is found between 50 and 90 km of altitude. It is difficult to study because radio soundings give no echoes there, nor can satellites be kept in orbit. This is because the density is still fairly high. Few electrons are formed. They undergo numerous collisions with neutral particles, causing a considerable absorption of the radio waves which should produce reflections.

Like the ionosphere as a whole, the D region changes a good deal from day to night. At night the electron density is virtually too small to measure; by day it is scarcely greater than 10^3 to 10^4 electrons/cm^3. The degree of ionization, defined by the ratio

$$\alpha = \frac{n^+}{n + n^+},$$

is very small (10^{-10}) in this region.

Immediately above comes the *E region*, which was the first to be discovered (Appleton). Figure 4 indicates the small, but important, place occupied by this region. It really exists only by day and is detected as a slight irregularity on the curve of electron densities. It is contained between the altitudes of 90 and 150 km, where electron densities of 10^5 electrons/cm^3 are found. At night this figure is reduced by a factor of 100. This region is characterized by great regularity in its behavior with respect to the Sun.

Beyond 150 km we enter the vast domain of the permanent and complex *F region*. It extends to great altitudes. Under certain conditions during the day, it splits and gives rise to two layers, called F_1 and F_2. The former is rather thin and does not extend beyond 200 km; the latter has a considerable thickness. The electron density can reach 2×10^6 electrons/cm^3 at altitudes of 300 or 400 km. Since the local density is low, the degree of ionization can reach 2×10^{-3}.

The behavior of the F_2 layer is characterized by numerous anomalies which make it a poor indicator of solar activity. We shall see that this is explained by the fact that the magnetic field must play an important role.

The part of the F region located above the maximum ionization level is not accessible to radio sounding. This means that our knowledge of the other side remained fragmentary until satellites could study it 'from above'. They have shown, among other things, that this layer extends much farther than predicted by the theories of some ten years ago.

It is still difficult to fix the upper limit. The electron density decreases less quickly than that of the neutral particles, so that the ionization coefficient rises. Since collisions between charged particles dominate, we have to consider a veritable plasma.* These collisions are very different from the all-or-nothing collisions which occur between neutral particles, or between neutral particles and electrons. In the present case the interactions are electric forces of attraction and repulsion, which operate at a distance.

As for the nature of the ions present, it is not surprising to learn that the percentage of helium nuclei (He^{++}) and protons (H^+) increases with height. With these components, the above-mentioned electric forces can become greater than the gravitational

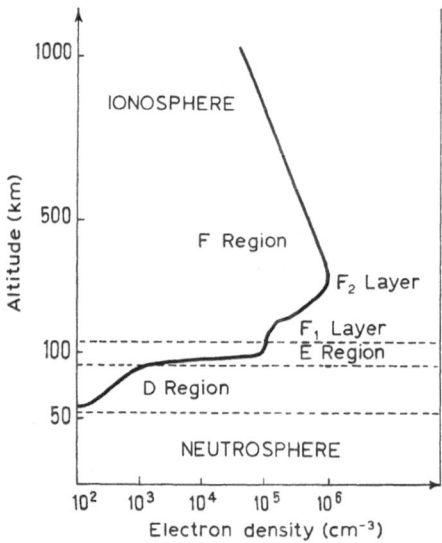

Fig. 4.   Typical vertical distribution of the electron density by day.

forces, whence there can arise new situations which we shall examine in due time. When protons become dominant, we say that we have entered the *protonosphere*.

The terrestrial magnetic field makes the medium very anisotropic, tending to modify the normal distribution of the ionization. The charged particles move preferentially along the lines of force, whose direction varies according to the latitude. In the polar regions, the lines of force are almost vertical; the diffusion of the charged particles is therefore of the same type as that of the neutral particles. The situation in the tropical zones is very different.

Another problem can be raised: that of the co-rotation of the medium with the Earth, which subjects the particles to large centrifugal forces and which effectively extends the ionized medium. However, as in the case of the neutral particles, there is a critical level. This is the level at which the density of charged particles is low enough

* Rigorously speaking, this term, borrowed from biology, designates a gas whose components have been ionized; however, it is often applied to a partially ionized gas.

for collisions and electrical interactions to cease to be effective. Then the magnetic field takes over, and we have reached the domain of the so-called Van Allen belts.

6. The Van Allen Belts

Going to still greater altitudes, we arrive in a region where the neutral particles have practically disappeared. Only electrons and protons remain. Their number is such that the medium remains electrically neutral.

These charged particles, whose presence was long suspected, are bound to the Earth not by the gravitational field but by the magnetic field, which forces them to remain at high altitudes. Because of their low density, interactions between them can be considered rare; their spatial distribution is principally an image of the terrestrial magnetic field, which will be described in the following chapter and which is characterized by its lines of force. Therefore this distribution depends not only on altitude, but also on magnetic latitude.

We shall see below that a charged particle placed in the terrestrial magnetic field is endowed with a complicated motion, causing it to move from one hemisphere to the other in the north–south direction while circling the globe much more slowly. For direct study, the instruments of nuclear physics (Geiger counters, photomultipliers, ionization chambers, etc.) are placed on board space vehicles.

Because of the structure of the terrestrial magnetic field, the charged particles avoid the polar regions and thus remain confined in zones between certain latitudes north and south, forming a vast toroidal belt around the Earth. In retrospect, it seems strange that the existence of this portion of the terrestrial environment was not known well before its actual discovery – which goes back only to 1958, as a result of the launching of the very first artificial satellites. Many theoretical works tended to predict its existence; but it is true that these works were concerned only with the fate of charged particles coming from outer space.

The discovery of this phenomenon is so recent that it is not yet possible to draw up a detailed map of the distribution of these charged particles as a function of their type (electrons or protons), of their energy, of altitude, or of the morphology of the magnetic field.

It has become customary to distinguish two large regions separated in altitude. They are called the *Van Allen belts*, after the American physicist who was the first to recognize their reality.

The lower one, called the *inner belt*, begins at an altitude of 600 km in the western hemisphere and 1600 km in the eastern hemisphere. Its base rises progressively as one approaches the equator. The same is true of its summit, which reaches an altitude of 9000 km. This belt is contained between $\pm 40°$ of latitude.

The second belt, called the *outer belt*, is much larger than the first. It extends between latitudes 55° and 65°. Its thickness is much greater (it is measured in tens of thousands of kilometers) (Figure 32).

This distinction now appears rather arbitrary; it was primarily due to the detection facilities offered by the first artificial satellites. In reality there is a smooth transition from one belt to the other. In the first belt protons are easily detected, while in the second it is mainly electrons that are counted.

Detection instruments register – with a certain efficiency – impacts on appropriate sensitive materials, whence there is a certain counting rate which must then be related to the physical characteristics of the medium (number of incident particles, energy and spatial distribution of these particles, etc.). The detector is more or less directional.

From the number of impacts registered, one must recover the corresponding flux.* This conversion problem is not easy. Particle energies are sorted out by absorbing screens of variable composition and thickness, which can be traversed only by particles whose energy is greater than a certain threshold. When photomultipliers are used, one must take into account the fact that electrons can give rise to X-rays by the inverse photoelectric effect, thus falsifying the measurements. Calibration can be performed on the ground with standard sources; but at present calibration is also performed in flight in order to increase the validity of the interpretation. It is obvious that the position of the satellite must be precisely known at all times, as well as its orientation with respect to the Sun and the Earth, and that the spacecraft must be prevented from turning around one of its axes. Great progress has been made in the last few years in apparatus for pointing (by means of solar and infrared sensors) and stabilization, so that more and more refined measurements can be made, even at very great distances from the Earth. But space is so vast, and the data collected at ground level so numerous, that the information obtained remains fragmentary.

Where do all these charged particles come from? It is not yet possible to give a complete answer to this question. First we must note that their motion in the terrestrial magnetic field cannot continue indefinitely. When they penetrate into the lower, and therefore denser, layers of the atmosphere, they undergo collisions with neutral particles which put an end to their career. They can also undergo interactions at higher altitudes which force them to change their velocity so that they are directed towards lower altitudes, which they would not have reached otherwise. Many other 'loss' mechanisms are possible.

Now, we know that the Van Allen belts exist permanently. It is true that they are

---

* The global flux is expressed in number of incident particles per unit surface area per unit time; but this concept is insufficient, for it does not distinguish between different energies and different directions. Depending on the detector used, one can determine:

– the unidirectional flux $J(E, \theta, \varphi)$, reduced to unit solid angle around the direction in question (θ, φ) and to unit energy interval. It is expressed in number of particles per square centimeter per second per steradian per mega-electron volt;

– the cumulative unidirectional flux, which is the result of integrating $J(E, \theta, \varphi)$ over the energy $J(>E)$;

– the unidirectional flux which results from an additional integration over solid angle, and therefore over the angles θ and φ.

Thus we write:

$$J(>E) = \int_E^\infty J(E)\, dE.$$

The 'spectrum' $J(E)$ is rather well represented by a relation of the form $J(E) = J_0 E^{-\gamma}$, where γ depends on the nature of the particles and on position in space.

subject to temporal variations that can depopulate or overpopulate the belts, but this phenomenon is temporary. Therefore one or more sources must exist. Various mechanisms have been imagined; none of them is entirely satisfactory. Let us merely say, for the present, that the only possible source is outer space.

7. The Frontier

As we continue to go farther from the ground level, we finally reach interplanetary space. Contrary to what one might imagine *a priori*, the transition is not smooth, for interplanetary space also has a structure different from that of a perfect vacuum. Between the Sun and the Earth there is a magnetic field arising in the Sun. Moreover, the Sun continuously emits a plasma now known as the *solar wind*. Only space explorations have really made it possible to understand how the connection is made.

The effect of the solar wind is to confine the terrestrial magnetic field to a cavity surrounding our globe, which the Earth carries along in its different motions. Beyond this cavity the Earth has practically no influence. Thus we have reached the limits of the terrestrial environment, which have a complex structure. It is to be expected that the frontier does not surround the Earth in a symmetrical fashion, since the Sun exercises a dominating influence. We must therefore distinguish between the sunlit hemisphere and the dark hemisphere, which are very dissimilar. Although the atmosphere in the broad meaning of the term ends at nearly 100 000 km from the Earth in the direction of the Sun, it extends farther in the opposite direction. Just exactly how far? It is not yet easy to be specific – but much farther than the distance indicated above.

In the years to come answers will certainly be found to these questions, thanks to the enormous budgets for space research authorized by the United States and the Soviet Union.

CHAPTER II

THE TERRESTRIAL MAGNETIC FIELD

It is obviously impossible to dissociate the effects of the three components of the terrestrial environment: gravity, magnetism, and matter.

The gravitational field does not present any particular difficulty. Its structure is simple. The weight of a body decreases as the inverse square of its distance from the center of the Earth. The gravitational field is practically unmodified by the presence of the atmosphere, whose mass is very small. Only the Sun and the Moon exercise perturbing forces, which make themselves felt in the atmospheric tides, about which we shall say a few words in connection with the study of the ionosphere.

The terrestrial magnetic field has a more complicated structure. Having already described it in a summary fashion, we shall go on in this chapter to consider its effect on the charged particles present above a certain altitude, as well as its influence on the propagation of electromagnetic waves, of which there are numerous sources both in man-made transmitters and in nature.

1. Description of the Earth's Magnetic Field

A. STUDY OF THE MAGNETIC FIELD

It was only in the fifteenth century that the distribution of the terrestrial magnetic field began to be seriously explored, thanks to the great maritime voyages. In 1701, the astronomer Halley drew up the first map of this field, and determined that it was well organized and underwent slow changes called secular variations. Since this period, measuring instruments have been much improved. Present-day maps, which are published regularly, are more and more precise, but are still not sufficiently accurate over the oceans and in the more inaccessible regions of the Earth.

As an overall description, we can say that in the first approximation the terrestrial magnetic field is that of a dipole, that is, that of an imaginary magnetized bar located in the interior of the Earth. This dipole has a *magnetic moment** $\mathcal{M} = 8.10 \times 10^{22}$ A m². Its axis does not coincide with the axis of rotation of the Earth, nor does its center coincide with the center of the globe. Figure 5 shows that the angular discrepancy is $11°4$ and that the geometric discrepancy is 400 km, in the direction of the Pacific.

Thus at each point of space there is a magnetic induction **B**, well determined in magnitude and direction. It is the sum of two terms:

– one, called the principal term, which has its origin – still poorly understood – in the interior of the globe and fluctuates only very slowly in the course of centuries;

* The moment of a bar is the product of one of its 'magnetic masses' and the distance between them.

– the other, much more agitated, which is due to currents induced in the ground by certain atmospheric phenomena.

There are regions where this field is notably different from that of the best theoretical dipole that can be found. This is the case in the South Atlantic, in Brazil, and near South Africa, where 'depressions' of 20% have been found.

If we go out to some distance from the Earth, we find that these anomalies are quickly weakened. They are attributed to localized internal currents, or to substantial magnetic deposits.

Using a mathematical technique known as 'spherical harmonic analysis', it is possible to calculate the distribution of the field far from the ground once that at ground level has been measured. This method requires the use of extensive computing facilities, and it improves as one uses larger and larger numbers of harmonics. With the first 512

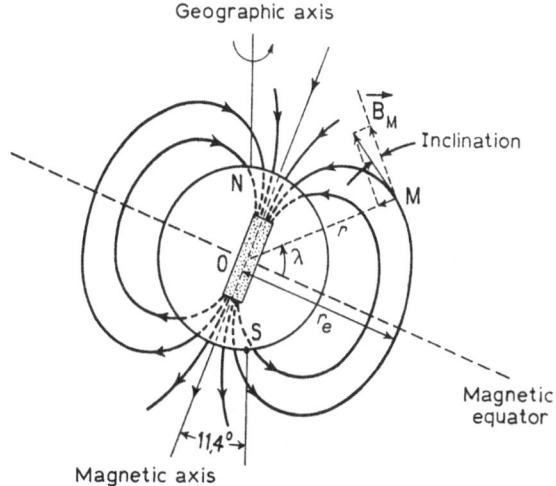

Fig. 5.   The terrestrial magnetic field and its lines of force (meridian section).

that are currently being used, obtained from 75 000 local measurements on the ground, one arrives at a fine structure comparable with the measurements made by space probes and satellites. But here we shall content ourselves with the principal term, represented by the dipole we have defined.

B. GEOMETRY OF THE MAGNETIC FIELD

Since the magnetic induction is a vector quantity $\mathbf{B}$, we must know three components in order to define its intensity and its direction. The dipole involves some simple calculations; we shall limit ourselves to stating the results.

Let us specify a point M in space (Figure 5) by means of its distance r from the center of the Earth and the angle λ between the direction OM and the plane which divides the dipole into equal parts. This plane intersects the Earth along a great circle called the *magnetic equator.* *

* This plane has scarcely any concrete physical significance. In reality, it is clear that the real magnetic equator is a closed, undulating curve that takes into account the surface irregularities of which we have spoken. But seen from space, this distinction is superfluous.

By symmetry, the vector $\mathbf{B}_M$ lies in the plane defined by M and the *magnetic axis NS*. This plane is called the magnetic meridian.

The two projections of $\mathbf{B}_M$ can be written (Figure 5):

$$B_r(M) = \frac{2\mathcal{M}}{r^3} \sin \lambda, \qquad B_\lambda(M) = \frac{\mathcal{M}}{r^3} \cos \lambda;$$

whence we obtain the intensity

$$|\mathbf{B}_M| = \frac{\mathcal{M}}{r^3} \sqrt{1 + 3\sin^2 \lambda} = B_{eq} \sqrt{1 + 3\sin^2 \lambda},$$

where B_{eq} is the value at the equator ($\lambda = 0$).

The terrestrial magnetic induction is weak. Expressed in gauss, it varies from 0.2 to 0.7 when one goes from the equator to the magnetic poles, values which are incomparably smaller than the induction to be found between the poles of a magnet. It decreases quickly (as $1/r^3$) with increasing altitude.

The preceding formulas are not very simple, since they depend on both distance (r) and direction (λ). It is of interest to have a panoramic view. For this reason we shall introduce the *lines of force*, imaginary curves defined as being everywhere tangent to the vector $\mathbf{B}$.

Figure 5 shows some of them, contained in a single magnetic meridian. They can be represented by the formula

$$r = r_e \cos^2 \lambda,$$

where r_e is the equatorial distance indicated on the same figure. It is clear that along such a line, the induction is not of constant intensity. The lines of force acquire a physical reality as soon as space is populated with electrically charged particles. Then we notice that the ionized particles tend to collect along these lines, and that certain radioelectric signals follow them under certain conditions. We shall see that they play an especially important role in the Van Allen belts.

Note that the magnetic induction is purely horizontal at the magnetic equator and vertical at the poles. Elsewhere, it has an *inclination* I with respect to the horizontal plane. This angle I is such that

$$\tan I = \frac{B_r}{B_\lambda} = 2 \tan \lambda.$$

2. Electrically Charged Particles and the Magnetic Field

A. CYCLOTRON MOTION

Since the space around the Earth is full of electrically charged particles, it is important to know what will happen to them in the magnetic field; what will become of those that may arrive from the Sun, and those that may be born on the spot from an ionization due to collisions or to solar radiation?

The force $\mathbf{F}$ exerted on a charge q in a magnetic field $\mathbf{B}$ is surprising at first glance for it acts in a direction perpendicular to **B** and to the velocity **v**. We bring out this fact by writing the force as a vector product:

$$\mathbf{F} = q\mathbf{v} \wedge \mathbf{B}.$$

The fundamental law of mechanics enables us to draw the following conclusions:
  - there is no acceleration in the direction of **B**;
  - the acceleration is perpendicular to the velocity;
  - this velocity is therefore constant in absolute value;
  - the energy of the charged particle does not change during its motion so long as the induction is not a function of time.

A more detailed study to be found elsewhere* shows that the charged particle q twists around **B** following a circular helix whose axis is a line of force (Figure 6), which in this way attains a kind of material existence.

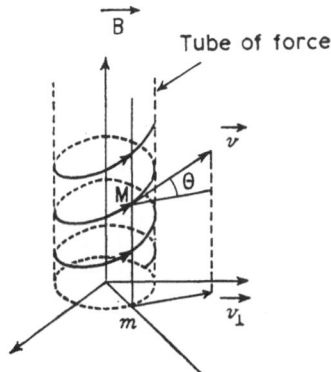

Fig. 6.   Motion of an electrically charged particle in a uniform magnetic field.

It is of interest to know how many 'spirals' are described per second. This quantity f_b is the *cyclotron frequency* or *gyrofrequency*

$$f_b = \frac{q\mathbf{B}}{2\pi m}.$$

It is known once we have determined the nature of the particle – that is, its mass m and its charge q – and the intensity $|\mathbf{B}|$ of the induction.

Thus a proton will have a cyclotron frequency nearly 2000 times smaller than that of an electron placed in the same position. With a typical value of several tenths of a gauss, we obtain values of several hundredths of a kilohertz for the electron – and much less if we go to a significant distance from the Earth.

As for the radius of the helix, called the *Larmor radius*, its value is:

$$\mathbf{R}_g = \frac{m v_\perp}{q\mathbf{B}}.$$

It depends not only on the nature of the particle, but also on its 'transverse' velocity

* Bibliography, p. 112.

$v_\perp$ which is represented in Figure 6. For convenience, we say that there corresponds to this velocity a 'transverse' energy $W_\perp = \frac{1}{2} mv_\perp^2$, so that the total, constant energy W of the particle can be considered as the sum of two terms: the energy $W_\perp$ and the longitudinal energy (in the direction of $\mathbf{B}$) $W_\parallel$.

In describing its spiral, the charged particle is equivalent to an electric current which generates its own magnetic induction. It behaves like a tiny magnet having a magnetic moment

$$\mu = -\frac{W_\perp}{B^2} \mathbf{B},$$

which opposes the induction $\mathbf{B}$.

Since W and $W_\parallel$ are constant, the energy $W_\perp$ is also constant by subtraction. For this reason the moment μ is called the first *adiabatic invariant*.

B. PARTICLE DRIFT

The charged particles found in the space surrounding the Earth are not located in a homogeneous magnetic field. They are, moreover, subject to gravity and are quite often located in electric fields. The simple helicoidal motion we have just studied is therefore only a first approximation.

A rigorous study of the various possible cases would be tedious. It is sufficient to know that whenever other forces are present, the particle is endowed with a *drift motion* at right angles to the induction $\mathbf{B}$ and to the resultant supplementary force.

Figure 7 gives an idea of this phenomenon. In the case of an electric field, the force is exerted in opposite directions for ions and electrons, so that the drift is the same. Ions and electrons move together in the same direction.

In the other cases (a force of mechanical origin or an inhomogeneity in the magnetic field), the drift depends on the sign of the charge, producing a separation between charged particles of different polarities and thus immediately introducing an electric field, which causes an additional drift.

We shall assume that the terrestrial magnetic field is not too inhomogeneous over a distance comparable to the radius of gyration. Such an assumption was made by Alfvén, who arrived at some important conclusions. Under these conditions one can in fact preserve the concept of a guiding center, a fictitious point which takes the particle along with it as does a hand that moves while holding a sling.

To ascertain that our hypothesis is valid, it is sufficient to verify the fact that the radius of gyration is always very small compared with that of the Earth, even when the charged particles have large velocities. Let us take, for example, a 10-MeV* proton at 3000 km from the ground. Its Larmor radius is only 3 km, and that of an electron of the same energy would be only forty times greater.

In the same way, an electrically charged particle will move in a complicated fashion in the terrestrial magnetic field. There will be three types of motion:

---

* The electron volt and its multiples (keV or kilo-electron volt, and MeV or mega-electron volt) are frequently used units. The electron volt is the energy acquired by an electron under a potential difference of 1 V. Thus we find that $1 \text{ eV} = 1.6 \times 10^{-19} \text{ C} \times 1 \text{ V} = 1.6 \times 10^{-19} \text{ J}$. This unit is very small, but it is on the right scale for the elementary particles.

(a) *Gyration*, which we have just described and which takes place at the frequency f_b, whose magnitude depends upon the local value of the induction.

– This frequency increases from the magnetic equator to the poles along any one line of force, as can easily be seen from the formulas given above.

– On the other hand, it decreases with altitude, as does **B**. It is possible to draw up a map of $f_b(\lambda, r)$ for each type of charged particle.

(b) *The north-south oscillation* of the guiding center along a line of force, and thus from one hemisphere to the other.

Let us try to explain this motion.

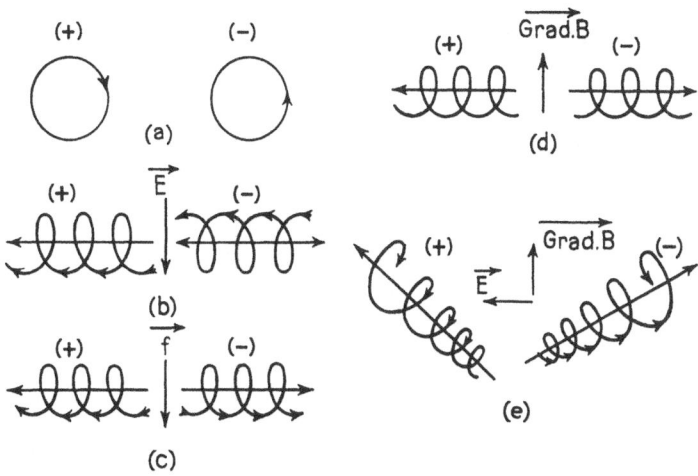

Fig. 7. Drift motions of electrically charged particles in a magnetic field under the influence of an additional force. (a) **F**=0. Simple gyration. (b) **F** is due to an electric field **E**. Protons and electrons drift in the same direction. (c) **F** is of mechanical origin. Protons and electrons drift in opposite directions. (d) The force **F** results from the fact that the induction **B** is not uniform. The drifts are again in opposite directions. (e) **F** is due to the simultaneous presence of an electric field **E** and an inhomogeneity in the magnetic induction. The 'spirals' become more open for both protons and electrons. The drifts are in opposite directions.

As we have said, the kinetic energy W of the charged particle remains constant, but now there is a force in the direction of **B**. It arises from the fact that the induction varies along the line of force. The velocity (and therefore the energy) along **B** is no longer constant. Since $W=W_{\parallel}+W_{\perp}$, the energy $W_{\perp}$ in the direction perpendicular to **B** changes also. And therein lies the new phenomenon.

To specify how the energies $W_{\parallel}$ and $W_{\perp}$ are exchanged in the course of the motion, it is sufficient to realize that the moment $|\mathbf{\mu}|$ of the particle remains invariant. Referring to Figure 6, which defines the angle θ between the velocity **v** and the induction **B**, we see that $v_1=v \sin \theta$; thus our condition $|\mathbf{\mu}|=$const. can be written

$$\frac{\sin^2 \theta}{B} = C_1 ,$$

where C_1 is a constant.

The physical significance of this formula is easy to understand.

Let us follow a line of force, starting at the equator. There the induction is a minimum, and so is θ. As we approach the poles, B increases and so does θ, the latter reaching a value of $\pi/2$. The charged particle has then reached the *mirror point*, from which it will begin to move in the opposite direction until it reaches the symmetrical mirror point in the other hemisphere, which is said to be magnetically *conjugate*.

Is there always a mirror point? We see that there is not, for it exists only if the induction is large enough, and this is not always the case. The mirror point may be 'virtual', that is, located in the interior of the Earth. It can also be located in the dense layers of the atmosphere, in which case the particle will behave quite differently upon reaching it.

The result of this analysis is that we can expect to see charged particles in the very high atmosphere with these motions, which will enable them to remain there for a very long time, until a collision (or a change in the magnetic field) stops them.

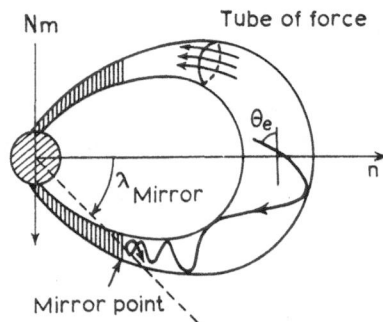

Fig. 8. North-south motion of a charged particle between mirror points.

The conservation of the magnetic moment μ means that the magnetic flux contained in the (nearly closed) spiral of gyration is invariant, and therefore that the charged particle remains on the surface of a single tube of force. This is shown in Figure 8.

The configuration of the terrestrial magnetic field is called a 'magnetic bottle', because of the constriction of the tubes of force. Other bottles on a much smaller scale are produced in laboratories, for they have the property of confining plasmas rather effectively.

(c) *The longitude drift of the guiding center*

This motion has two principal causes:

– the centrifugal force felt by the center as it follows a curved line of force;

– a variation of the terrestrial magnetic field perpendicular to the line of force (Figure 7d).

Thus the charged particles will go around the Earth while performing the general type of motion outlined in Figure 9. For clarity, the helicoidal motion has been omitted.

Ions and electrons drift around the globe in opposite directions.

The time-scales of these different motions are very different. The oscillations in latitude have periods that are easy to determine from the length of the lines of force and the field strength. They are measured in seconds. On the other hand, the motion in longitude is slow. The electrons make one turn around the Earth in a time which depends upon the altitude, the mirror latitude, and their energy. This can be a matter of minutes or of hours, depending on the case.

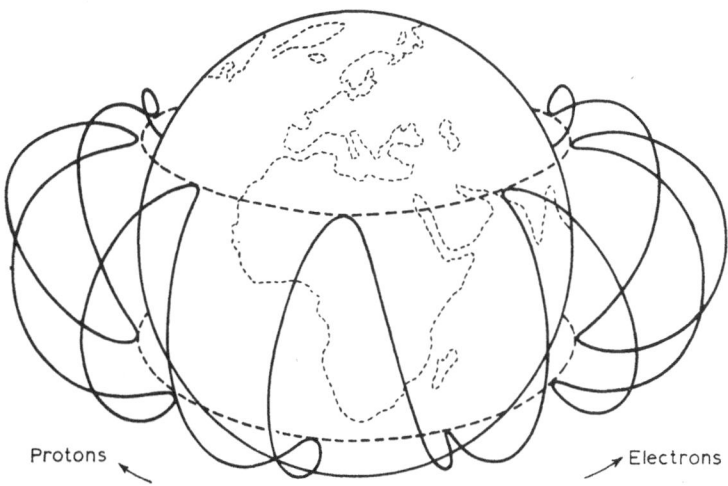

Protons   Electrons

Fig. 9. Longitude drift of electrically charged particles around the Earth (according to J. O'Brien).

C. BEHAVIOR IN THE REAL MAGNETIC FIELD

The preceding discussion was true for a dipole, but the description continues to be applicable in the true geometry of the magnetic field.

In reality, the lines of force are twisted and do not lie in a plane, and their symmetry vanishes not only around the terrestrial magnetic axis but also between the two hemispheres. New situations result.

In order to understand them, let us dwell a little longer on the case of the dipole. Each particle, characterized by its energy W and its constant moment $|\mu|$, oscillates between mirror points. These are the points where the induction $|B_r|$ is just sufficient to reflect that particle.

Where are these mirror points?

They are obviously located on the surface $|B_r| = \text{const}$, whence we derive the equation:

$$\frac{\mathcal{M}}{r^3} \sqrt{1 + 3\sin^2 \lambda} = \text{const}.$$

Such a surface completely surrounds the Earth (Figure 10).

Thus our charged particle $(W; |\mu|)$ remains on an *invariant surface* which is obtained by rotating a line of force about the terrestrial magnetic axis, and which is bounded by its intersection with the preceding surface.

What becomes of this arrangement in the real case?

The charged particle $(W; |\mu|)$ will undoubtedly still describe a surface bounded by the mirror points, but will it be a closed surface? Will the particle return to the line from which it started?

Northrop and Teller* have answered these questions by introducing a second, so-called 'longitudinal' invariant of the motion. If it is effectively constant, the charged particle will remain on a fixed surface which depends on its energy, its *two* invariants, and its starting point. After going once around the Earth, it will be back in the same position.

Now let us consider two charged particles which have the same energy W and the same invariant $|\mu|$, but which start from two different points on the same line of force. They will be together again on this line after one complete revolution but not in between, for it is only after a complete revolution that they will have experienced equal drifts in longitude. They describe the same surface, but not in the same way.

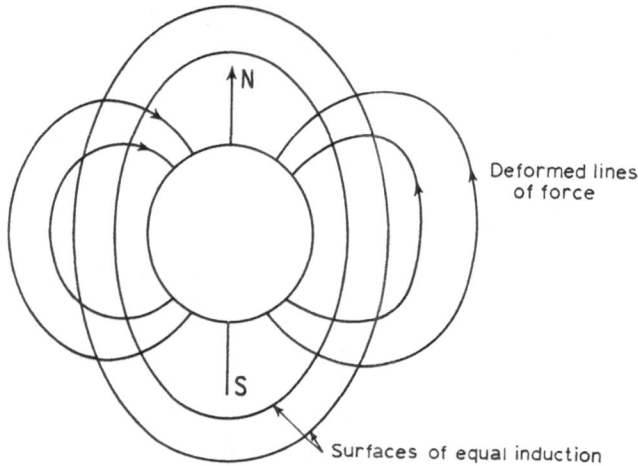

Fig. 10.

If, on the other hand, the energies are different, the surfaces described will be different, but will intersect in the starting line.

All these rather general results became of interest when the Russians and the Americans produced nuclear explosions at very high altitudes, liberating a very large number of electrons born on the same line of force, but rapidly sweeping out different invariant surfaces only to come together again on the original line. The layer they form should be thinnest in the neighborhood of this line.

D. STÖRMER: A PRECURSOR

We have just seen the confining effect of the terrestrial magnetic field. Now, the Sun emits charged particles which react with this field when they arrive in the vicinity of the Earth. We would like to know how they behave. The problem has not yet been solved for a distribution of charged particles, but Störmer treated it – at the price of a relatively laborious mathematical development – in the case of individual corpuscles.

The discovery of the magnetosphere made this work very timely, and many labora-

* Bibliography, p. 112.

tories are now trying to simulate the behavior of charged particles projected towards magnetized spheres.

Starting from the basic equation already cited, •

$$m \frac{d\mathbf{V}}{dt} = q\mathbf{v} \wedge \mathbf{B},$$

and adopting the dipole model, one can obtain 'first integrals' of the motion of a particle of charge q and mass m. They are written

$$\left(\frac{dr}{ds}\right)^2 + \left(\frac{d\varphi}{ds}\right)^2 + \left(\frac{dz}{ds}\right)^2 = 1,$$

$$r^2 \frac{d\varphi}{ds} = \frac{r^2}{\rho^2} + 2\gamma,$$

where r, φ, and z are the coordinates of the particle as indicated in Figure 11, s is the curvilinear abscissa of the particle on its trajectory, and γ is a certain constant.

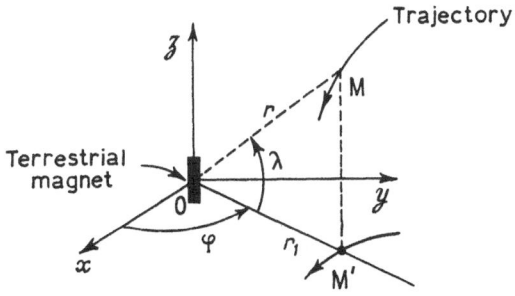

Fig. 11.

The first relation only expresses what we already know: the energy of the particle remains constant throughout its motion.

The second introduces the constant γ, which remains to be specified.

If the problem is to be solved completely without the aid of a computer (which Störmer did not have), one needs a third equation, as yet undiscovered. But the first two equations are sufficient for us.

Let us consider only the case in which the particle remains in the plane of the magnetic equator, $x0y$ in Figure 11. It is located by its coordinates φ and r_1, and describes a trajectory $r_1(\varphi)$.

Störmer showed that there are permitted regions in this plane, and other regions which are forbidden to the particle. These regions are not pre-existing portions of space, but depend on the charged particle itself and its velocity. Calculation shows that, in order to represent the phenomenon conveniently, it is advantageous to express the distances involved not in kilometers but in terms of a unit called the Störmer unit, whose value is:

$$C_{st} = \sqrt{\frac{\mu_0 \mathcal{M} q}{4\pi \, mv}}.$$

This unit is very large: 9 billion kilometers for 1 keV electrons or 0.5 eV protons, and 3 million kilometers for very high-energy electrons and protons.

The permitted zones are then defined by the two inequalities:

$$\gamma > -\frac{1+r_1^2}{2r_1}\,; \qquad \gamma < \frac{r_1^2-1}{2r_1}\,.$$

We have drawn in Figure 12 the two hyperbolas $\gamma = -(1+r_1^2)/2r_1$ and $\gamma = (r_1^2-1)/2r_1$ which bound these zones. Only the white surfaces are permitted.

Consider a charged particle which arrives in the vicinity of the Earth. There is a particular value of γ associated with it. On the graph, this particle follows a path parallel to the axis of ordinates until it reaches P_1, where it is reflected to return towards

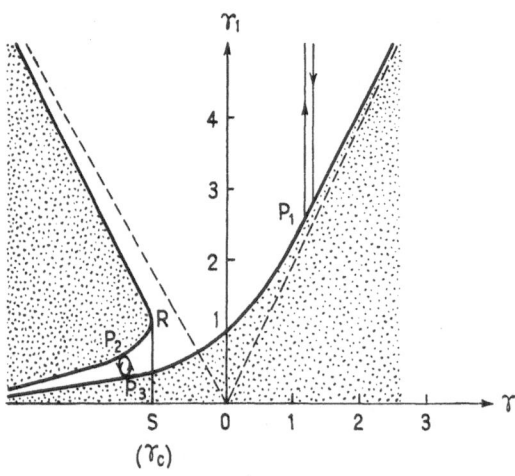

Fig. 12. A charged particle from outside cannot reach the white region to the left of RS, but a charged particle that is already in this region will remain there and oscillate from P_2 to P_3 ($\gamma < \gamma_c$). A charged particle from outside is reflected at P_1 and sent far from the Earth.

outer space. The bar RS at the 'critical' abscissa γ_c plays an important role, for the white region to the left of it can never be reached from outside. If, on the other hand, a particle is already in this region, it will remain there, oscillating between the points P_2 and P_3.

In other planes the conclusions are the same, but the shape of the permitted zones is different.

This theory could have made it possible to predict the existence of the Van Allen belts. It should be modified to take into account our new knowledge of the terrestrial magnetic field far from the Earth and of the solar emission, in order to understand the interactions which are the basis of such phenomena as magnetic storms.

3. Electromagnetic Waves in an Ionized Medium and in the Presence of the Earth's Magnetic Field

We have already seen that the atmosphere is ionized to a greater or lesser extent above a certain altitude, and that radio soundings make it possible to obtain information concerning the state of the medium. With a view to the study of the ionosphere, it is of interest to examine the behavior of an electromagnetic wave under such conditions.

We shall first consider the simplest case, that in which the charged particles are entirely free. They are then subject only to a single force, that due to the electric field **E** of the wave arising from a natural or artificial transmitter. The expression for this force is a simple one: $\mathbf{F} = q\mathbf{E}$, where q is the charge. Under these conditions, there is no transfer of energy from the wave to the particles.

We shall then introduce collisions, which hinder the preceding motions and result in an absorption of energy as soon as the particle density becomes significant. These effects are expressed by an additional force, called a force of friction.

Finally, we must take the terrestrial magnetic field into account. Its presence is felt in a third force, with which we are already familiar. Thus we arrive at the so-called magneto-ionic theory, which is so complex that we can only begin to attack it.

A. THE COLLISIONLESS MEDIUM

Since ions and electrons have very different masses, one can always neglect the motion of the former where the atmosphere is concerned, at least in a first approximation. The ions then serve only to maintain the overall electrical neutrality of the plasma.

An electron of mass m and charge q takes on an oscillatory motion at the frequency f of the wave characterized by its periodic electric field $\mathbf{E} = \mathbf{E}_0\, e^{i\omega t}$; this is equivalent to a current $q\mathbf{v}$, where $\mathbf{v}$ is the velocity of the electron. This velocity is easy to calculate, for $m\,(d\mathbf{v}/dt) = q\mathbf{E}$. If we have N electrons per unit volume, the current becomes $Nq\mathbf{v}$. It is added to the Maxwell displacement current, and modifies the index of refraction of the medium, which becomes

$$n^2 = 1 - X \quad \text{with} \quad X = \left(\frac{f_0}{f}\right)^2 .$$

We have defined $f_0^2 = (Nq^2)/(4n^2 m\varepsilon_0)$. This parameter f_0, which has the dimensions of a frequency, is characteristic of the plasma (ε_0 is the permittivity of free space). For this reason it is called the *plasma frequency*. Introducing numerical values for the constants we find:

$$n^2 = 1 - 8 \times 10^{-5} \frac{N}{f^2} .$$

The index of refraction, which is practically equal to unity in the neutral gas, has become smaller. It depends on the frequency. We say that there is dispersion. The index obviously goes to zero when $X = 1$, that is, when $f = f_0$ or

$$N_0 = 1.23 \times 10^{-2} f_0^2 \text{ electrons/m}^3 .$$

If our transmitter works on a frequency $f > f_0$, the index is less than 1, but real. The wave traverses the medium. In the opposite case, the index no longer exists. The wave does not penetrate, but is reflected. For this reason, the plasma frequency f_0 is called the *critical frequency*.

B. THE MEDIUM WITH COLLISIONS

The components of the atmosphere undergo collisions. So long as the medium consists mainly of neutral particles, the collisions which occur in practice are between electrons and neutral particles.

It is, of course, both uninteresting and impossible to follow the individual history of each electron. Fortunately, the kinetic theory of gases enables us to define a typical electron, which leads to the desired result. It is as though only such 'mean' electrons existed, all behaving in the same manner. The typical electron is characterized by its velocity between collisions and by the number of collisions ν which it undergoes per unit time. This number ν depends on the particle density and on the temperature of the medium.

What is new with respect to the preceding case is that although all the electrons are 'typical', they do not vibrate in synchronization, for they do not all undergo collisions at the same time. Again, kinetic theory makes it possible to take this effect into account. We shall not go into detail. It is sufficient to know that everything takes place as if the electron were subject to a force of friction proportional to its velocity, to its mass, and to the number of collisions ν.

The index of the medium takes on a more complicated form:

$$n^2 = 1 - \frac{X}{1+z} - i\,\frac{Xz}{1+z^2},$$

with $z = \nu/2\pi f$.

It has become 'imaginary', expressing the fact that the wave is progressively absorbed as it passes through the medium. Moreover, this absorption also depends on the frequency.

Very diverse situations can be encountered, depending on the respective values of ν, f, and f_0. Two of them are clear:

– if $z \ll 1$, that is, if the medium has a low density (ν small) or if the frequency of the wave is high, we practically recover the collisionless case;

– if $z \gg 1$, that is, if the medium is heavily populated or if f is small, the absorption will be large. Even if the wave can pass, it will be quickly attenuated. If not, it will be reflected like light from polished metal.

C. THE REAL CASE

The terrestrial magnetic field subjects the electron to an additional force $q\,(\mathbf{v} \wedge \mathbf{B})$ with which we are already familiar, so that the equation of motion becomes

$$m\,(d\mathbf{v}/dt) + m\nu\mathbf{v} = q\,(\mathbf{E} + \mathbf{v} \wedge \mathbf{B}).$$

The electrons have to combine a helicoidal motion with an oscillatory motion which is interrupted at each collision, from which there arises a rather complicated situation.

The angle φ between **B** and the direction of propagation of the wave plays an important role. Appleton and Hartree have shown that the index conforms to the following relation (which is named after them):

$$n^2 = 1 - \frac{X\,(U - X)}{U\,(U - X) - Y_\perp^2/2 \pm \sqrt{(Y_\perp^2/4) + Y_\parallel^2\,(U - X)^2}},$$

where

$$Y_\perp = \frac{f_b}{f}\sin\varphi, \qquad U = 1 - iz,$$

$$Y_\parallel = \frac{f_b}{f}\cos\varphi, \qquad f_b = \frac{qB}{2\pi m}.$$

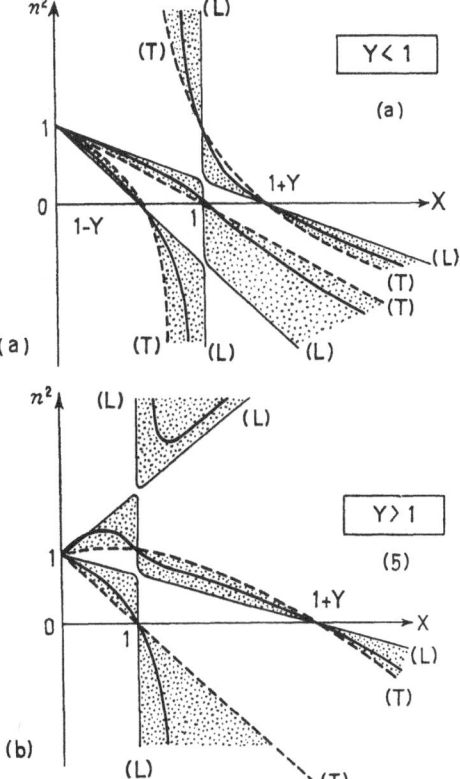

Fig. 13. Index of refraction of a magnetized plasma with respect to a wave of variable frequency. The intermediate curves correspond to a particular value $|0 < \varphi < \pi/2.|$

It is out of the question for us to discuss this expression here. It is represented graphically in Figure 13, in the case of $Y < 1$ and in the case of $Y > 1$ – that is, for a gyrofrequency less than and greater than the frequency of the wave. Most of the time, we encounter the case of $Y < 1$.

The important fact from our point of view is that there are now two values of the index, as a consequence of the + and − signs appearing in front of the radical. The medium has become birefringent. To one incident ray coming from the source, there correspond two rays in the ionized medium.

When the angle φ is zero, the wave propagates in the direction of **B**; this is the *longitudinal* case. When $\varphi = 90°$, it propagates in a direction perpendicular to **B**; this is the *transverse* case. As φ varies from 0 to $\pi/2$, the curves $n^2(X)$ in Figure 13 sweep out the hatched regions bounded by the curves marked (L) and (T), respectively.

Let us simplify the Appleton–Hartree relation as much as possible by assuming that $\varphi = 90°$ and $z = 0$ (or $U = 1$), which means that the propagation is transverse and that absorption is entirely negligible. We are left with

$$n^2 = 1 - X, \qquad n^2 = 1 - \frac{X(1-X)}{1 - X - Y^2} \qquad \left(Y = \frac{f_b}{f}\right),$$

that is, the dashed curves in Figure 13a. These curves intersect the axis of abscissas at three points, having abscissas 1, $1+Y$, and $1-Y$, and corresponding to the critical frequencies.

The first of these, f_0, is analogous to the result obtained without induction. It is called the *ordinary* critical frequency.

The second, f_x, is called by contrast the *extraordinary* critical frequency.

The difference between them, $f_x - f_0$, is equal to $2f_b$, as can be proved by a very simple calculation. This quantity provides a method of measuring the Earth's magnetic field at any altitude, when f_x and f_0 are determined by radio sounding from the Earth.

D. THE WHISTLER MODE

Let us consider the case in which collisions are negligible and the wave remains close to the magnetic lines of force. The index becomes:

$$n^2 = 1 - \frac{X}{1 \pm Y \cos \varphi},$$

Let us make a further simplification by considering a frequency which is very small compared to the gyrofrequency. There remains

$$n^2 = \pm \frac{X}{Y \cos \varphi},$$

and therefore only a single mode (the + sign). This is the ordinary mode illustrated in Figure 13b.

Such a case is applicable to the lightning discharges which are natural sources of radio waves with frequencies distributed over the whole spectrum. The lowest of these frequencies propagate while remaining close to the lines of force; they penetrate the exosphere, and can be used to study it.

This phenomenon was first pointed out by Barkausen in 1919; he was able to receive directly in an amplifier frequencies of several kilohertz – audible frequencies.

It was Storey, in 1953, who first gave a satisfactory explanation of such a reception. The signals emitted by the storm at very low frequencies can circulate from one hemisphere to the other and return, after reflection. Thus one hears the successive echoes, separated by the travel time.

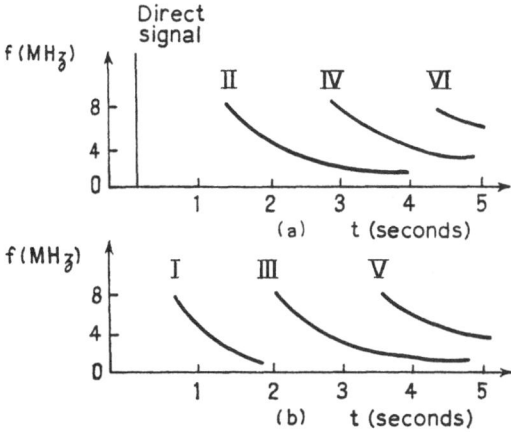

Fig. 14.   Dispersion law of whistlers at reception. (a) The reception takes place in the vicinity of the emitting discharge. Besides the original signal, one receives signals that have made 1, 2, 3 ... round trips along the line of force. (b) The reception takes place in the zone magnetically conjugate to the emitting discharge. There is no direct signal, and one receives signals that have made 1, 3, 5, ... one-way trips.

In practice, the velocity of the waves is a function of their frequency. This *dispersion* spreads out the reception in time, with the lower frequencies arriving later; hence the sound slides from high to low, and these signals have been given the name of *whistlers*.

When we calculate the time necessary to go to the other hemisphere, we find the expression

$$t = \frac{1}{2c\sqrt{f}} \int \frac{f_0}{\sqrt{f_b}} \, ds ,$$

where ds is directed along the line of force, f is the frequency of the wave, f_b is the gyrofrequency, f_0 the plasma frequency all along the trajectory, and c the velocity of light (Figure 14).

The wave spends most of its time in the upper part of the exosphere, where the induction (and therefore f_b) is smallest. A measurement $t(f)$ will thus give information concerning regions of higher altitude if one 'listens' at higher latitudes (Figure 15).

The above equation is only approximate, and is not applicable to whistlers which

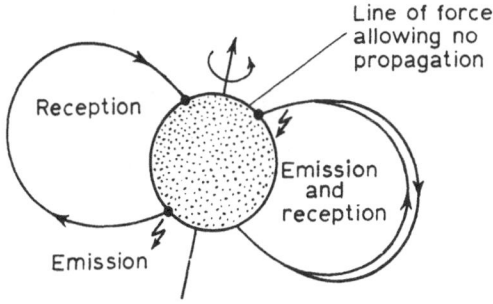

Fig. 15.   'Even' and 'odd' propagation along the lines of force.

reach very great altitudes, for the gyrofrequency becomes too small. For these waves, the curve $t(f)$ has a characteristic 'nose' shape. A high and a low sound are received simultaneously.

E. HYDROMAGNETIC WAVES

When the frequency of the waves becomes especially small, the ions have time to move during a period, and their influence can no longer be neglected. The index of refraction takes on an additional ionic term, and can be expressed:

$$n^2 = 1 - \frac{X}{1+Y} - \frac{(m_e/m_i)\,X}{1 \mp (m_e/m_i)\,Y} .$$

Since m_e is much smaller than m_i, this can be written

$$n^2 = 1 + \frac{X}{(m_e/m_i)\,Y^2} ,$$

at least for the ordinary mode.

The velocity of the wave in the plasma can easily be deduced from this expression, for the index is the ratio between the velocity of waves in a vacuum and in the plasma. The latter is designated V_A from the name of the Swedish physicist and Nobel prize winner Alfvèn, and its value is

$$V_A^2 = \frac{C^2}{1 + [X/(m_e/m_i)\,Y^2]} .$$

Replacing X and Y by their previously defined values, we find that

$$V_A = \frac{B}{\sqrt{\mu_0\rho}} ,$$

where ρ is the density of the plasma and μ_0 the permeability of free space.

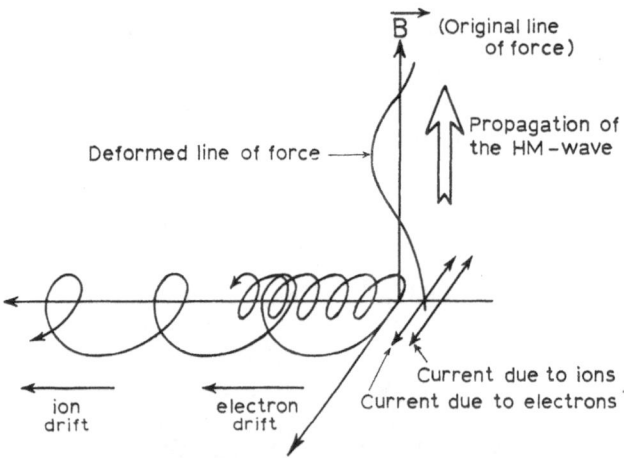

Fig. 16.　Schematic representation of a hydromagnetic wave. (The frequency is much less than the gyrofrequency of the ions and, *a fortiori*, than that of the electrons.)

This velocity is much less than the velocity of light, and it no longer depends on the frequency of the wave.

To get a physical picture of the behavior of hydromagnetic waves, it is sufficient to note that the charged particles are subject to the electric field **E** of the wave, which varies very little during the time required for the electrons and ions to perform a large number of gyrations in the magnetic field. These charged particles will have a drift velocity perpendicular to the fields **E** and **B**, and in the same direction for all particles. After a certain time, this drift will change direction, as does the field of the wave. No current is created.

Along the line of **E**, the situation is different, for ions and electrons move in opposite directions.* To this current there corresponds an induction which is added to the induction of the medium, and which results in a deformation of the lines of force along the path of the wave. The lines of force act like vibrating strings, and the hydromagnetic waves appear like vibrations of tubes of force carrying plasma (Figure 16).

---

* A current is a displacement of electric charge.

CHAPTER III

THE TERRESTRIAL IONOSPHERE

The ionosphere begins at an altitude of about 50 km and extends to beyond 1000 km. It is formed by interactions between the components of the atmosphere and the solar radiation. Let us begin by examining the characteristics of the latter. We shall then consider the various processes it causes and their consequences with regard to the Earth's atmosphere.

1. Solar Radiation

A. THE SUN AND ITS LUMINOUS RADIATION

When we look at the surface of the Sun with the naked eye, using neutral filters to reduce its light, we perceive a disk with sharp edges: the disk of the *photosphere*. It is occulted during total eclipses of the Sun. But this is a very brief phenomenon, quite rare at any particular place, and often rendered unobservable by atmospheric conditions. Since the invention of the coronograph by B. Lyot, eclipse conditions can be reproduced artificially.

Two other regions can then be distinguished:
– the *chromosphere*, ring-shaped and very thin;
– the *corona*, only faintly luminous but very extended.

The photospheric disk, with its diameter of 1.4 million kilometers, is 110 times as large as the Earth. Most of the solar mass is situated below the photosphere. This mass is 330 000 times the mass of our planet. A small calculation shows that the density is only 1.4, as opposed to 5.5 for the Earth.

Thus the Sun is formed principally of light elements. Ionized hydrogen is the major constituent. It enters into sustained nuclear reactions which compensate for the energy lost through radiation. The radiated energy is spread over the entire spectrum according to the well-known Planck distribution, at least in the first approximation.

Before we were able to observe the Sun from outside the atmosphere, only two bands of the spectrum were known to us.

The first, in the visible region, begins rather abruptly at 2900 Å * and ends at around 8000 Å. It corresponds to the intensity of a black body – or a state of thermodynamic equilibrium – at about 6000 K.

The second region is very different, extending from 1 cm to around 15 m. This is the range of solar radio astronomy (Figure 17).

Since it became possible to obtain the whole of the spectrum by means of rockets, it has been recognized that the black-body law is not faithfully reproduced. We can

* The ångström unit equals 10^{-10} m.

speak only of equivalent temperatures in given wavelength regions. Thus in the ultraviolet, the temperature is only 5000 K. Moreover, there are many emission lines and absorption lines, the latter being characteristic of the media traversed by the radiation.

The luminous radiation is emitted by molecules and by neutral and ionized atoms. It originates in the photosphere.

The chromosphere, which has a thickness of 10 000 km, is transparent to the underlying radiation. Its temperature is higher than that of the photosphere, ranging from 5000 K at its base to 50 000 K towards its summit.

The edge of the chromosphere is disturbed. Constantly changing jets of matter are observed there; they fall back after having attained considerable altitudes.

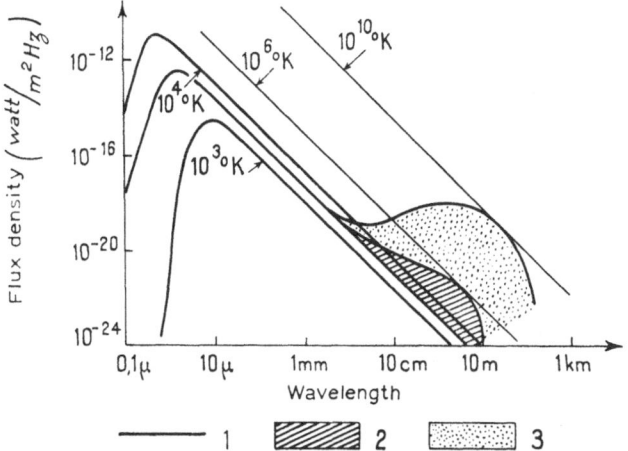

Fig. 17.   The solar spectrum and a black-body spectrum for various temperatures. (1) Thermal emission. (2) Non-thermal emission from the quiet Sun.
(3) Non-thermal emission from the active Sun.

As for the corona, it constitutes almost all of the solar atmosphere. Its density decreases as the temperature increases from 50 000 K to 2 million degrees; the temperature then falls back to 50 000 K at the distance of the Earth's orbit. It is in the corona that radio waves originate, as the result of various interactions between fields and particles.

Towards short wavelengths, the photospheric intensity decreases rapidly and the chromospheric and coronal emission dominates. Strong lines exist in the far ultra-violet. They play an essential part in the ionization of the terrestrial atmosphere. We note in particular the Lyman α line of hydrogen (1215 Å) and the resonance lines of neutral and ionized helium (584 and 304 Å, respectively).

Rockets and satellites have opened even the X-ray region (2 Å) to our inspection; it shows a large number of lines, not all of which have been identified as yet. When the total energy flux in the visible region – and thus of photospheric origin – is measured, it is evident that this flux varies only slightly from a mean value called the *solar constant*. However, the energy contained in the part of the spectrum below 100 Å is

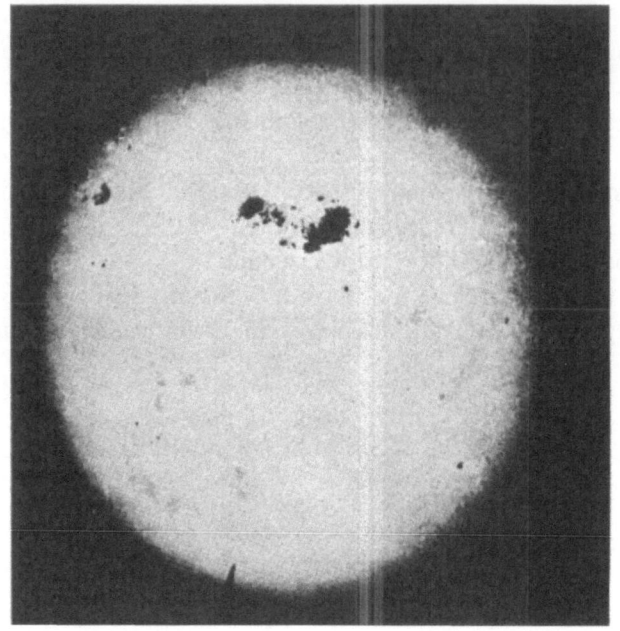

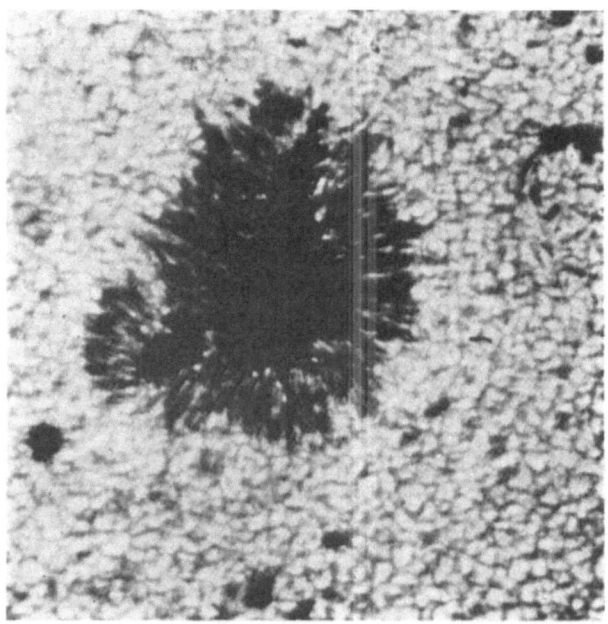

Fig. 18.   (a) Groups of sunspots. (*Princeton Observatory photo.*)
       (b) Structure of a sunspot. (*Princeton Observatory photo.*)

subject to fluctuations as great as a factor of seven. X-ray photographs show that the emission of the disk is far from being evenly distributed, even in a quiet period.

B. THE SOLAR CYCLE

The surface of the Sun is rarely homogenous, even in the visible region. Dark spots are observed there, sometimes so extended that they can be made out with the naked eye. Observatories have been counting them for more than two centuries. Figure 18 shows groups of spots of various extent, surrounded by lighter filamentary structures and floating on regions brighter than the rest of the disk, which have been given the name of *plages*.

In 1849, R. Wolf introduced the solar number R, defined by the empirical relation

$$R = k\,(10g + f)\,,$$

where k is a factor appropriate to the observer, f the number of individual spots, and g the number of groups of spots. Although it is rather arbitrary, this *Wolf number*

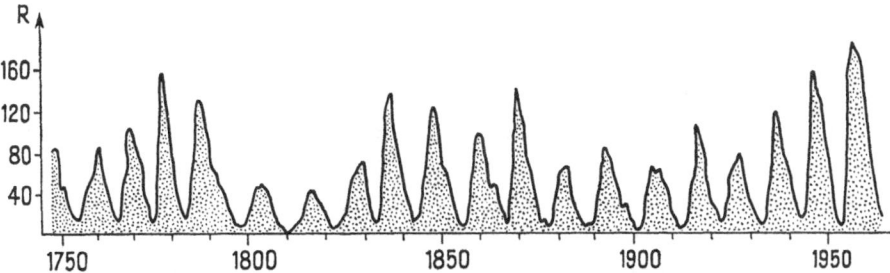

Fig. 19.   Variation of the Wolf number over two centuries.

gives a good description of the state of the Sun with respect to many terrestrial phenomena.

The Wolf number undergoes rather irregular periodic variations, with a period of eleven years. At the time called the 'solar minimum', R is practically zero. There are neither spots nor groups of spots. Shortly afterwards, the spots begin to appear on both sides of the equator at latitudes of around 40°. Some of them can be followed for several days, or even for several weeks. They are carried along by the Sun's rotation on its axis, which has a period of twenty-seven days.

As the *solar cycle* progresses, other spots appear, but at lower and lower latitudes. The last of them are formed 8° from the equator. Figure 19 shows the variation of the number R over the last two centuries. We see immediately that there are weak cycles and strong cycles, which appear to follow each other with a period of seventy-eight years.

The temperature of the spots is lower than that of the surrounding regions. In addition, the spots are the site of strong magnetic fields. By measuring the Zeeman effect* in absorption lines, we find intensities of several thousand gauss. Moreover,

* A magnetic field causes each line emitted to be split into several components located on either side of the original line, which is weakened. The separation of the components is proportional to the induction.

each spot has a north or south magnetic polarity. In any one group, one of these polarities dominates, and it has been shown that the groups are paired. To each of them, there corresponds another group of opposite polarity which is located at the same latitude, but about 10° distant in longitude. Because of these fields, a net transfer of energy takes place above the spots, leading to a relative cooling. This convection also takes place in normal regions, as is indicated by the granular structure whose bright points indicate rising columns of hot gas, while the darker points correspond to cool descending columns.

The cyclical variation of R is accompanied by an inversion of the magnetic polarity of the pairs of spots. If the leading spot in a group in one hemisphere has north polarity for eleven years, it will have south polarity during the following cycle. Therefore the period is really twenty-two years. It is not yet possible to know whether the Sun is subject to slower fluctuations of this sort, having a period longer than several human generations.

C. ATMOSPHERIC ABSORPTION

Let us imagine a beam of light from the Sun impinging upon the terrestrial atmosphere. As it penetrates the atmosphere, it will lose photons and therefore intensity. These photons disappear by transferring their energy to the components of the atmosphere. The molecules become dissociated and the atoms are either excited or ionized. These phenomena are called photoexcitation and photoionization, respectively.

Upon its arrival, the beam is composed of photons of all energies or (what amounts to the same thing) of radiation of all wavelengths. The depopulation takes place selectively. Certain photons are absorbed in large numbers, others little or not at all. Thus there are spectral regions of great transparency called atmospheric windows, which connect us with the outside world and which we have already cited. Let us now examine the regions of high absorption. There are two principal regions of this type:

– the first extends in wavelength from 1 cm to 0.7 μ, and thus over both the radio region and the infrared. Its absorption spectrum consists of bands due to transitions between energy states of molecules,* principally O_2 and N_2. These are symmetric molecules, and therefore contribute little to the absorption at long wavelengths.

Nitrogen scarcely absorbs at all, and oxygen does so only weakly. However, since O_2 exists in large quantities, the absorption in the red is significant. Compounds such as CO, CO_2, N_2O, etc. play the most important part, despite their very low concentration.

– The second absorbing region is of more interest to us. At the ground, we observe that it begins at around 3500 Å as a result of the presence of ozone at higher altitudes.

---

* Each molecule and each atom is characterized by various possible states of internal energy. Their normal state corresponds to the lowest energy, a situation which naturally occurs very often. In order to change them to a different state, energy has to be added. In the present case, this energy is in the form of photons. Very schematically, a transition from an energy state E_1 to an energy state E_2, where $E_2 > E_1$, takes place according to the relation $E_2 = E_1 + h\nu$, where $h\nu$ is the energy of the photon.

The ozone molecule (O_3) results from the association of three oxygen atoms and is produced by triple collisions of the type

$$O + O_2 + N \rightarrow O_3 + M \, .$$

It is not necessary to specify the nature of the molecule M which remains unmodified. This process requires the presence of oxygen atoms. The dissociation of O_2 is easily produced by ultraviolet radiation, either between 2000 and 2400 Å (the Herzberg band) or between 1500 and 1700 Å (the Schuman–Runge band). Since the penetration of solar radiation is very different in these two bands, there will be two ozone production zones, one at around 100 km and the other at around 25 km. The latter is more

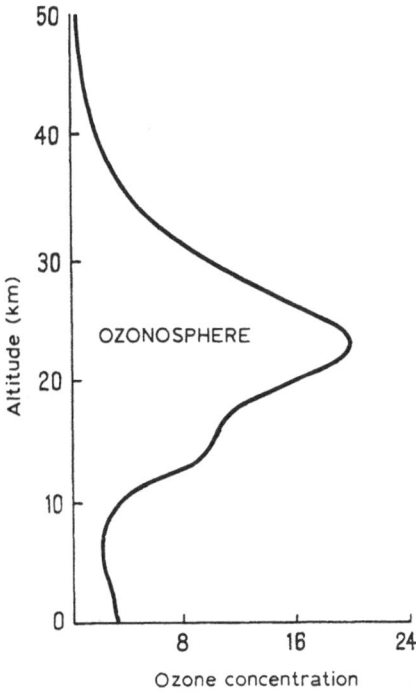

Fig. 20.   Distribution of ozone in the atmosphere.

important, for at this altitude there is a significant probability of triple collisions. Therefore there is an *ozonosphere* whose altitude and thickness depend on place and time, which persists during the night, and which forms an indispensable screen for organic life (Figure 20). When one rises to an altitude of 65 km, the effect of the ozonosphere becomes negligible and the spectrum is revealed down to 2000 Å.

As we have already said, there is total absorption at ground level for wavelengths less than 2900 Å. In going to higher altitudes, the absorption diminishes and a more extended spectrum is received.

Any absorption process can be characterized by a *penetration depth*. This is the distance the radiation must travel in order to be reduced by a factor of $e=2.7$ (the base of the neperian logarithm system), for the phenomenon is exponential.*

Figure 21 shows the altitudes at which only a fraction $1/e$ of the incident radiation is received. It indicates the constituents that are responsible for the absorption.

Between 2400 and 2000 Å, O_2 absorbs, but only weakly. This band is favorable for the production of ozone, of which we have already spoken.

Below 2000 Å, absorption by O_2 becomes more effective. It reaches a maximum at around 1500 Å. We see in Figure 21 that down to 1000 Å the curve shows irregularities, due to a superposition of the discrete absorption bands on the O_2 continuum. Thus

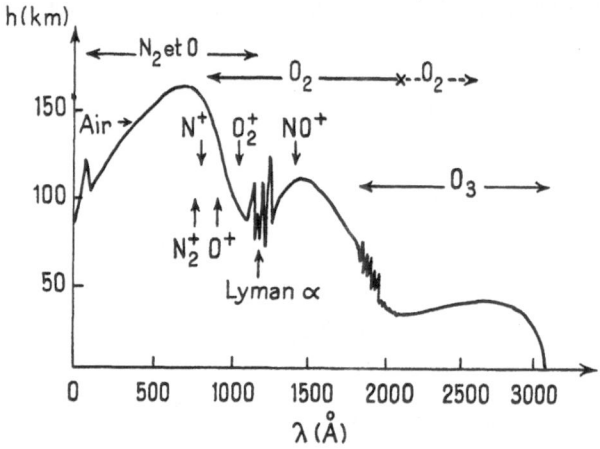

Fig. 21.—Altitudes at which the solar radiation is reduced by a factor of $e=2.7$ (after Nawrocki, Watanabe, and Smith).

there are gaps reaching to within 70 km of the ground. One of them – the most pronounced – is due to the Lyman α line of hydrogen.† In particular, radiation of this wavelength ionizes the NO molecule and plays an important part in the formation of the lower ionosphere, as does X-ray radiation. At wavelengths lower than 1000 Å,

_____

* If I_0 is the incident intensity, the transmitted intensity I will be only

$$I = I_0 \exp\left(-\sum_i \alpha_i \frac{n_i}{n_0}\right),$$

where n_0 is the number of neutral molecules per unit volume at ground level, α_i is the absorption coefficient per unit length for the component of type i, and n_i is the number of molecules of type i per unit volume. We see that n_i depends on the altitude (cf. Chapter I).

† The hydrogen atom has only one orbital electron, which can gravitate around the nucleus on different paths fixed by the laws of quantum mechanics. The atom is in the fundamental state when the electron moves in the path closest to the nucleus. Any 'excitation' from outside forces it into a different path. In the inverse process, the atom returns to the state of minimum energy by emitting a photon whose frequency can be written

$$\nu = A \left(\frac{1}{n^2} - \frac{1}{m^2}\right),$$

where A is a fundamental constant, and n and m are integers which represent the quantum number of the final and original orbit, respectively.

The Lyman series corresponds to $n=1$, with m greater than 1. When $m=2$, we have the Lyman α line.

it becomes possible to ionize the molecular (O_2 and N_2) and atomic (O and N) components arising from dissociations. The maximum wavelengths corresponding to these processes are 1026, 796, 910, and 852 Å, respectively. The nitrogen atom is produced by a recombination of the molecular ion N_2^+, a very high-altitude process. Photodissociation requires a wavelength shorter than 1270 Å. Only the Lyman α line penetrates far enough to produce atomic nitrogen at moderate altitudes.

2. Formation of the Ionosphere

All the absorption processes we have just reviewed produce a profound modification in the atmosphere, and this modification depends on the altitude. The most important consequence is the production of ions and electrons.

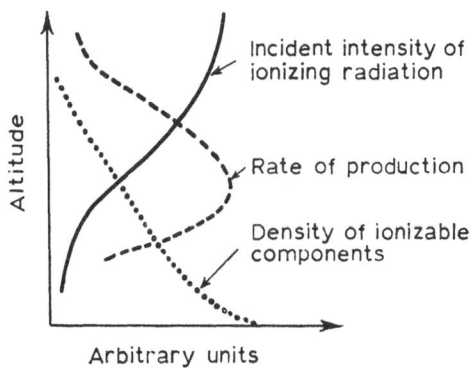

Fig. 22. Production of ionization.

Using a simplified theory due to Chapman, we can get an idea of the overall profile of the ionosphere. This theory rests on a great number of simplifications. It assumes:
– that the photoionizing radiation is monochromatic;
– that there is only one ionizable component;
– that the charged particles disappear at the same point where they are created;
– that there is only one loss mechanism: recombination;
– that the density decreases according to a simple exponential law;
– that the temperature is unique.

Let us designate by I_0 the intensity of the incident beam of mono-energetic photons. I (h) represents the number of photons per second striking a surface of unit area at altitude h. In general, the beam will be inclined at an angle χ to the vertical, for the Sun is not at the zenith. After having traversed the thickness dh (traveling a distance $dh/\cos \chi$), this beam will have lost an intensity (Figure 23)

$$dI = A\rho I \left(\frac{dh}{\cos \chi}\right),$$

where ρ is the density of the atmosphere.

Electrons are produced by this absorption at the rate of

$$q = \beta \left(\frac{dI}{dh}\right) \cos \chi$$

per unit time per unit volume.

Since $\rho = \rho_0 e^{-h/H}$, we can calculate $I\,(h)$, dI/dh, and $q\,(h)$. We find the optimum production of electrons at an altitude

$$h_m = H \log_e \left(\frac{A\rho_0}{\cos \chi}\right),$$

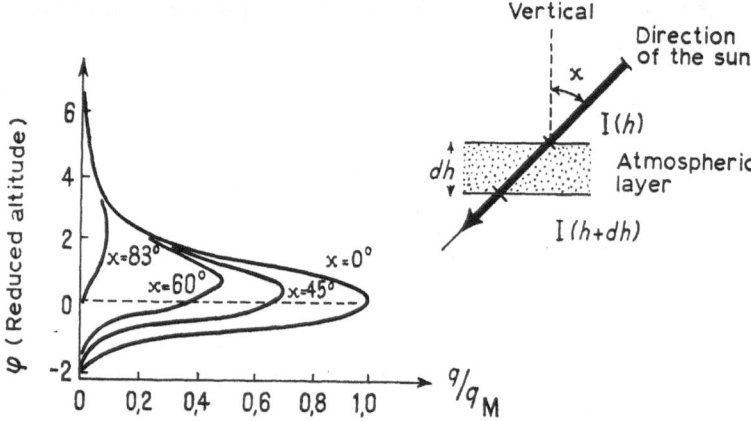

Fig. 23. Profile of the Chapman model, as a function of the inclination of the rays of the Sun.

and this rate of production is

$$q_m = \frac{\beta \cos \chi I_\infty}{He} = q_M \cos \chi,$$

where e is the base of the neperian logarithm system.

At an arbitrary height h we will have

$$q\,(h) = q_M \exp \left(1 - \varphi - \frac{e^{-\varphi}}{\cos \chi}\right),$$

where φ is the *reduced* height, measured in units of the scale height:

$$\varphi = \frac{h - h_m}{H}.$$

It is interesting to trace the ratio q/q_M as a function of φ, for various values of the angle χ. We obtain a profile of the ionosphere (Figure 23) which bulges more and more as the Sun approaches the zenith. Since we have assumed that the ionized particles disappear by simple, on-the-spot recombination, the electron population will change at the simple rate $dn/dt = q - \alpha n^2$, where α is called the recombination coefficient.

In particular, in the equilibrium state in which the Sun produces as many electrons as disappear, we will have at all altitudes

$$n_e(h) = \sqrt{\frac{q(h)}{\alpha}} \quad \text{electrons.}$$

There is an obvious asymmetry between the ionization above and below the privileged altitude h_m. This result is to be expected, for the density increases as we approach the ground.

The Chapman model gives only a very preliminary idea of the real situation, particularly as regards the thickness of the layer. More advanced theories have made it possible to come closer to the truth, but none of them is entirely satisfactory as yet. Many additional phenomena have to be taken into account. For example:

– the real composition of the atmosphere;

– the temperature (it is obvious that the dissociations and ionizations will result in a temperature increase, since recombinations take place only as a result of various collisions between neutral particles, ions, and electrons);

– the real spectral composition of the solar radiation;

– the sphericity of the Earth;

– the movement of the charged particles, set in motion by the winds and more or less guided by the terrestrial magnetic field;

– the other possible mechanisms for the loss of charged particles, in addition to recombination (the electrons can become attached to neutral particles, the particles can exchange charge during collisions, etc.).

3. The Lower Ionosphere

The Chapman theory predicts the existence of only a single layer. Considering the defects of this theory, it is not surprising to discover that there are actually several layers. Moreover, it is customary to distinguish between regions and layers of the ionosphere. In principle, the regions are the result of a particular physical process. The layers are stratifications of a single region.

It is also customary to distinguish between the lower and upper ionospheres. The former includes two regions called D and E. The latter consists of the F region. Because of its extent in altitude, the F region might seem at first glance to correspond to Chapman's analysis, the lower regions then being local irregularities. In fact, this is not the case. Despite its small thickness, the lower ionosphere is a rather faithful representation of Chapman's characteristics. The F region, on the other hand, presents many anomalies which are only beginning to find a valid explanation.

A. THE D REGION

This region is located between 60 and 85 km, in a medium whose composition remains similar to that of air at ground level while the temperature varies from 130 to 250 K. Its presence explains the strong absorption experienced by radio waves during the daytime if they are reflected at higher altitudes, for collisions between electrons and neutral particles are very frequent in the D region.

Electrons are relatively rare in this zone: $10/cm^3$ at 60 km, several thousand at the summit. The limits are difficult to specify, for it is not easy to make a direct study. Soundings from the ground give no echoes, and satellites cannot be maintained in orbit at these altitudes. We are left with rockets, and certain methods like partial reflection due to the scattering of electromagnetic waves from ionization irregularities.

Two formation mechanisms are to be invoked:

– the Lyman α line reaches the D region, and can ionize NO molecules there. Although this constituent is rare, it is sufficient to liberate electrons in acceptable numbers;

– the O_2 and N_2 molecules can liberate electrons, especially during certain periods of the solar cycle.

The seasonal variations of the electron density are poorly known. A great regularity has been observed from one day to another in summer, but not in winter. This winter anomaly appears as an intensification of the layer at high latitudes, where the Sun is less effective because of its large zenith angle χ.

The D region plays an important part in the transmission of radio waves between stations on the ground, by causing the strong absorption of which we have already spoken; this absorption is all the more pronounced the lower the frequency (cf. Chapter II). This explains why the reception of stations operating on medium wavelengths is poor during the daytime, especially in summer. Communications specialists define a minimum usable frequency which depends on the path-length, the geographic region, the time, and the power of the transmitter.

For long wavelengths, on the other hand, the D region behaves like a metallic mirror that improves communications, which are also effected by a ground wave. The Earth and the D region thus form a veritable wave guide.

B. THE E REGION

Located immediately above the preceding zone, this region has a higher degree of ionization. It can be studied by the method of radio sounding.

We know that an ionized medium is characterized by a plasma frequency $f_0 = Nq^2/4\pi m\varepsilon_0$, which depends on the electron density N. There is such a frequency for every altitude. The highest frequency corresponds to the maximum electron density of the region. It is called the *critical frequency* f_E. By making soundings at increasing frequencies f we will therefore observe:

– a reflection of the wave at a level lower than that of maximum ionization (so long as $f < f_E$);

– a reflection at the maximum level for $f = f_E$;

– a transmission with no return if $f > f_E$.

In this way we form traces $f(h)$ called *ionograms*. Figure 24 shows one of them. The different regions appear in succession.

The traces are double because of the terrestrial magnetic field. We must therefore distinguish between the ordinary and extraordinary critical frequencies. The altitudes

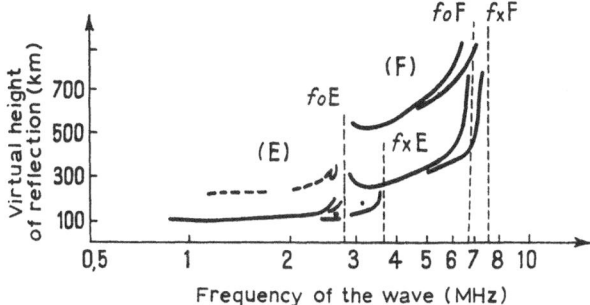

Fig. 24. Ionogram obtained by vertical sounding at high latitudes. ———— Ordinary and extraordinary reflections. Note the echoes corresponding to two round trips between the ground and the ionized layer.

h indicated are virtual heights. Physically, they represent only the travel time of the wave in the course of its round-trip journey. If we suppose that the velocity remains that of the waves in a vacuum, we obtain the indicated graduation. In reality, the velocity of the wave depends on the state of the medium through which it travels, and it is of interest to convert these heights into real altitudes. This operation is mathematically laborious but very important to carry out, if one is to take real advantage of the information contained in the ionograms. After it has been performed, the profiles obtained are much smoother (Figure 25).

The E region is situated between 85 and 130 km. The electron density can reach $10^5/cm^3$. The corresponding critical frequency is several megahertz. According to

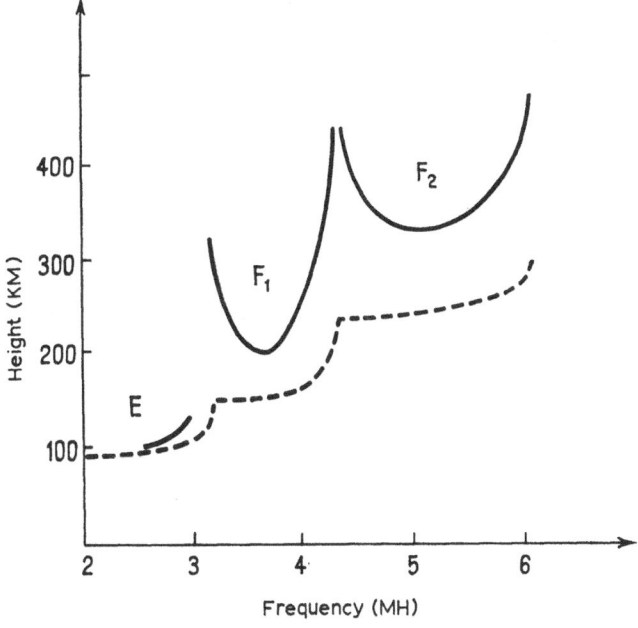

Fig. 25. Observed ionogram and true ionogram. ———— Virtual heights.
— — — — — Real heights.

Chapman's theory, it should depend upon the position of the Sun in the local sky through the relation

$$f_0 E = C_E (\cos \chi)^{1/4},$$

which explicitly expresses the effects connected with χ (time, season and place). The factor C_E, on the other hand, contains the influence of the state of the Sun.

This expression is verified by the observations, enabling us to conclude that this region 'follows' the Sun.

However, there are still some departures from Chapman's law. They have been largely taken into account by allowing for the variation of the scale height with altitude. The exponent of $\cos \chi$ is then modified.

Changes due to the state of the Sun are well represented by a formula of the type

$$(f_0 E)^2 = a (1 + bR),$$

where R is the Wolf number. The coefficient b is small, but since R varies from 0 to 150 during a cycle, the effect is noticeable.

The correlation with abnormal solar emission in the X-ray and ultraviolet regions is even better. We may therefore suppose that this radiation is responsible for the existence of the region.

Recent studies by rocket have shown that very short wavelengths (2 to 8 Å and 33.7 Å) ionize O_2 and N_2 at the base of the region. The ionized molecules then give rise to NO^+, which is the predominant ion. Higher up, between 95 and 115 km, the ionization appears to be due to radiation in the extreme ultraviolet, and the O_2^+ ion then becomes as abundant as the NO^+ ion.

At night, the ionization decreases in the absence of the Sun. But its value can still be measured by special sounders operating at low frequencies.

4. The High Ionosphere

The upper ionosphere is composed of a single region, called the F region. It begins at about 140 km. It has a considerable thickness, so that the D and E regions appear only as bumps on the overall ionization profile.

The upper limit has not been precisely fixed. It depends on the criterion chosen. On the other hand, the level h_m of maximum ionization is well defined. It can vary from 250 to 500 km.

We shall consider three zones in succession:
– the base of the region;
– the ionization located below h_m;
– the ionization located above h_m.

This distinction arises from the fact that only the part located below h_m is accessible to radio sounding. The waves whose frequency would enable them to be reflected higher up have already been reflected from the lower levels. It has only recently become possible to study the upper side by putting equipment on board satellites, of which the best known is *Alouette*.

A. THE F_1 LAYER

By day, the F region often divides into two layers called F_1 and F_2. In such cases, the F_1 layer appears a little after dawn and reaches its maximum development a little after noon. It is not always possible to define a critical frequency – that is, an ionization maximum. The layer is then seen only as an excess of ionization on the edge of F_2. When the critical frequency $f_0 F_1$ exists, it is found to satisfy a relation of the form

$$f_0 F_1 = C_{F_1} (\cos \chi)^{1/5} ,$$

showing the controlling influence of the Sun in the sense predicted by simple theories like that of Chapman. The exponent depends on the latitude and C_{F_1} contains the state of the Sun, as for the E region.

B. THE LOWER F_2 LAYER

The fundamental characteristic of the F region is that it behaves in a much more complicated fashion than the preceding regions. It is marked by many 'anomalies', or profound departures from the simple theories.

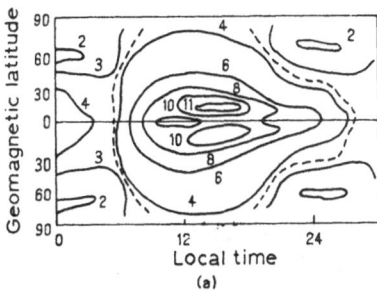 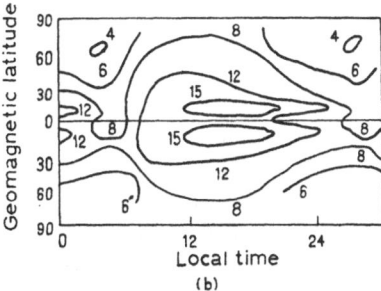

Fig. 26. Equidensity profiles of the F region at the equinox. (a) Minimum of the solar cycle. (b) Maximum of the solar cycle. The numbers represent the critical frequencies in MHz.

Figure 26 shows the ionization distribution at the maximum and at the minimum of the solar cycle. An examination of the iso-electronic contours provokes the following remarks:

– the controlling influence of the Sun is obvious. The ionization increases rapidly as soon as the Sun rises, but the maximum is rarely reached at noon. In some parts of the Earth, it is reached before noon, and in other parts later. The nocturnal decrease is often slow and a secondary maximum is sometimes observed during the night;

– the equinoxes do not produce a clear symmetry about the equator. When the graphs are redrawn as a function of latitude or, better still, of magnetic inclination, a definite improvement in the symmetry is observed;*

– the maximum values of $f_0 F_2$ are not found on the equator – not even on the magnetic equator – but along two belts located on either side of the magnetic equator at 10 or 15° of latitude. Thus there is an equatorial depression;

* It is the magnetic latitude or magnetic inclination as they are measured at ground level that must be introduced here, for the altitude is relatively low. The dipole approximation is too rough for the ionosphere.

– the diurnal variation of the ionization is greatly affected by the terrestrial magnetic field;

– the seasonal variation is also baffling, for the ionization is maximum in winter, especially during periods of intense solar activity;

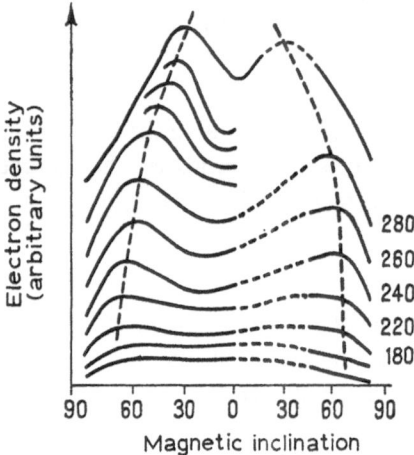

Fig. 27.   Ionization profiles as a function of the magnetic inclination at various altitudes. The magnetic anomaly appears beyond 200 km (after S. A. Crom, *Nature*, 1959).

– at polar latitudes, there is still a diurnal variation although the atmosphere remains light or dark for long periods of time;

– at night, the ionization is greatest in the equatorial zone and it is not very sensitive to the time, the season, or the phase of the solar cycle.

Before attempting to explain these results, we must note that the altitude of maximum ionization varies greatly with time and with the magnetic conditions. It is therefore preferable to determine the electron density N (h) at specified altitudes as a function of inclination. Presented in this way, the facts become more comprehensible. Diagrams of this type can be made only when one knows the real height at which the

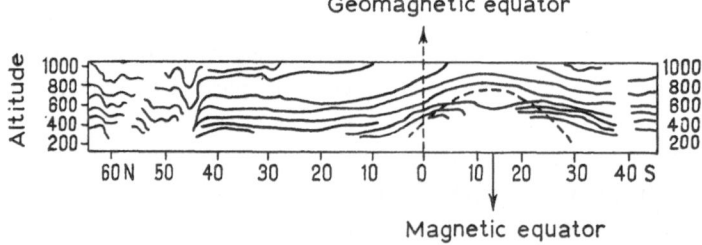

Fig. 28.   Lines of equal electron density in a magnetic meridian plane.

waves are reflected. This requires, as we have said, a rather laborious mathematical treatment, but it is performed more or less routinely at the present time.

Then we observe that the profiles are normal at the F_1 level, but that higher up the magnetic anomaly appears in the form of two symmetric humps (Figures 26 and 27) which approach the magnetic equator as the altitude increases. This effect continues above the level of maximum ionization and the two humps finally come together, as has been proved by satellite measurements.

The behavior of the ionization in the F region cannot conform to Chapman's theory. One must take into account important phenomena which this theory neglects from the very beginning. Among them, we particularly note:

– the change in the composition of the neutral gas with altitude. At the base of the region, we find mostly O^+ and N_2. Higher up, the medium becomes richer in atomic oxygen and poorer in molecular nitrogen. Helium and hydrogen appear next. The scale height therefore increases considerably;

– the mechanisms for producing electrons, which are no longer due only to photo-ionization, but also to detachment from negative ions during collisions;

– the disappearance of electrons through several processes, for direct recombination becomes less important than attachment to molecules (O_2) and 'electronegative' atoms (O);

– the displacement of the whole mass of charged particles during their lifetime, which is extended by the low density of the neutral particles. There are many causes for these motions: electromagnetic forces, temperature gradients, solar and lunar tides, diffusion in the gravitational field.

These facts can be formally taken into account by writing an equation for the change in the electron density of the form

$$\frac{dn_e}{dt} = q(h, t) - D - M,$$

where:

– $q(h, t)$, the rate of formation of the charged particles, remains the same as in Chapman's theory except for monochromaticity;

– D, the loss term, is mostly due to an exchange of charge between ions and molecules, followed by a dissociative recombination;*

– M is a motion term which increases with altitude, and is due to three effects: pressure and temperature variations, and the terrestrial magnetic field.

It is not possible to discuss all these phenomena in detail. Moreover, taking them into consideration does not enable us at the present time to account for all the anomalies of the F region. However, they do have the advantage of explaining the existence of these anomalies.

5. Dynamics of the Ionosphere

The ionosphere is the site of coherent motions which involve the neutral particles, the ions, and the electrons. The neutral particles move under the influence of variations in temperature, pressure, and gravity. The others are carried along with the neutral gas when it is dense enough. This is the case in the D region. Beginning with the E

---

* The exchange of charge takes place between an oxygen ion and a neutral molecule: $O^+ + M \rightarrow M^+ + O$, where M is either O_2 or N_2. The dissociative recombination involves an electron and the ion M^+: $M^+ + e \rightarrow M'^* + O^*$, where the ion M^+ contains oxygen.

region, the influence of the terrestrial magnetic field becomes more and more important. Above a certain altitude, it becomes the most important factor, as we shall see in the next chapter. At the level of the F region, it is already indispensable for an explanation of the anomalies.

A. MOTIONS OF THE NEUTRAL ATMOSPHERE

These motions can be conveniently studied up to an altitude of about a hundred kilometers by observing meteor trails. Similar trails are now regularly produced by rockets which release metallic vapors, such as sodium vapor. The observed direction of the winds is zonal – that is, east–west with seasonal variations in magnitude as well as in direction.

Superimposed on, and often more important than, these prevailing winds, are the tidal phenomena, which play an important role. The oceanic tides are well known. They are due to the combined attraction of the Sun and the Moon, which modulates the terrestrial gravitational field. The same is true for the atmosphere, and the tides are easily observed by analyzing the behavior of meteor trails or the critical frequencies of the E region, for example.

At low latitudes, the diurnal component (24 hours) dominates, but as one goes away from the equator the semi-diurnal component (12 hours) emerges more and more clearly. The existence of this semi-diurnal tide has been known at ground level for more than two centuries through the analysis of barometer readings. For a long time it was attributed to the attraction of the Sun, but it is now thought to be principally a thermal effect, for the Moon, which produces the same gravitational perturbations as the Sun, does not give rise to a similarly large phenomenon.

If the tide is mostly due to the heating of the atmosphere by the Sun, one might ask why its period is 12 hours rather than 24 hours. There is probably a resonance effect involved, but it is not yet well understood. In any case, it is experimentally determined and theoretically confirmed that the amplitude of the tidal wave increases greatly with height. It is 1000 times greater at 100 km than at ground level.

Finally, there are also irregular motions, again established by the analysis of meteor trails and too systematic to be due to turbulence.

The ionization irregularities at the level of the E layer provide another method of studying the circulation of the neutral gas, which is still dense enough to carry the ionized particles along with it. A signal at fixed frequency is recorded by three receivers separated by several wavelengths. It is then easy to deduce the magnitude and direction of the motions of the irregularities. Sometimes they change their shape as they move, as clouds do. It then becomes difficult to reach an unambiguous conclusion. Diurnal components of 12 and 24 hours are found, in agreement with those given by meteor trails; there are also seasonal effects that depend on the geographic position.

B. MOTIONS OF THE IONIZED ATMOSPHERE

The altitude displacements of the neutral gas, whether they are of thermal origin or are due to tidal phenomena, have important consequences for the motion of the charged particles.

These motions are of interest for two reasons:

– they enable us to predict the existence of systems of ionospheric currents on a global scale at the level of the E region, producing the periodic variations in the terrestrial magnetic field which are measured at observatories;

– they redistribute the ionization vertically as well as in latitude, explaining certain aspects of the behavior of the F region.

The charged particles carried along at a velocity **u** by the neutral gas form a moving conductor in the terrestrial magnetic field. In this conductor there arises an induced current, as in a dynamo. This is the 'dynamo effect' characterized by the existence of the field $\mathbf{E} = \mathbf{u} \wedge \mathbf{B}$.

The ions and electrons therefore move with respect to the neutral gas with a velocity **v**, which can be calculated by applying the following additional forces:

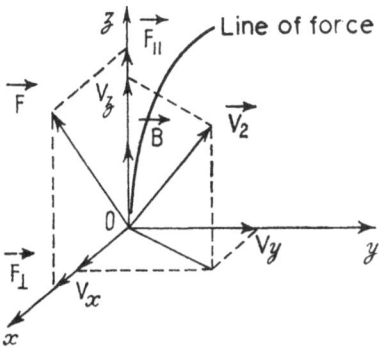

Fig. 29.

– a force of friction proportional to the frequency of collisions with neutral particles;

– the total Laplace force $q(\mathbf{u}+\mathbf{v}) \wedge \mathbf{B}$, since their velocity relative to the Earth is $\mathbf{u}+\mathbf{v}$;

– a force of electrical origin $q\mathbf{E}$, where **E** is an electric field which arises from the separation of the charges.

The situation is complicated, and we shall give only a few rough ideas of it.

When we calculate the velocity v_e of a typical electron which undergoes v_e collisions per second and is subject to the force **F**, we obtain components which can be written (in the coordinate system displayed in Figure 29):

$$v_x = \frac{F_\perp}{m_e} \frac{v_e}{v_e^2 + \omega_{be}^2}, \qquad v_y = -\frac{F_\perp}{m_e} \frac{\omega_{ge}}{v_e^2 + \omega_{be}^2}, \qquad v_z = \frac{F_\parallel}{m_e} \frac{1}{v_e}.$$

Similar expressions can be found for the ions.

These components are unequal in magnitude, so that the conduction of the ionosphere depends strongly on direction. Thus we recover the anisotropy introduced by the terrestrial magnetic field.

However, these formulae can be simplified in the ionosphere, for the frequency of collisions ν_e is always less than the gyrofrequency ω_{be} (they are equal at about 70 km). Then we write:

$$v_x = 0, \qquad v_y = -\frac{F_\perp}{m_e}\frac{1}{\omega_{be}}, \qquad v_z = \frac{F_\parallel}{m_e}\frac{1}{\nu_e}.$$

The drift velocity v_y is independent of the number of collisions, and therefore of the altitude; the ions move at the same velocity as the electrons along the lines of force, and this velocity increases with altitude. This phenomenon is called *ambipolar* diffusion.

If we neglect pressure forces and gravity, $\mathbf{F}$ is reduced to $q_e\,(\mathbf{E}+\mathbf{u}\wedge\mathbf{B})$. The expression in parentheses is the total electric field $\mathbf{E}_t$. The components become

$$v_x = 0, \qquad v_y = -\frac{E_{tx}}{B}, \qquad v_z = \frac{E_{tz}}{B}\frac{\omega_{be}}{\nu_e}.$$

The ions and electrons have an identical drift normal to the electric field and to the

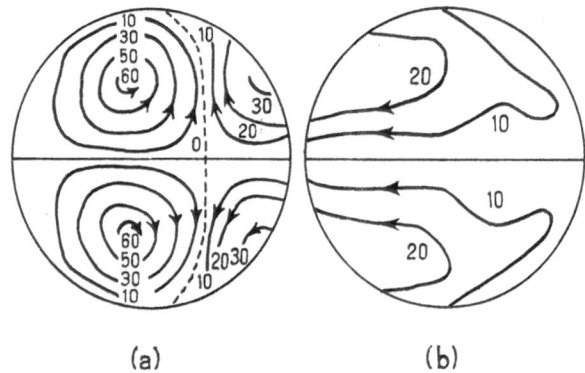

(a)   (b)

Fig. 30. System of electric currents circulating at the level of the E region and responsible for the diurnal variations in the terrestrial magnetic field (Chapman and Bartels). (a) Sunlit hemisphere, at the equinox and the minimum of the solar cycle. (b) By night, at the equinox and the minimum of the solar cycle.

terrestrial magnetic field; they move along $\mathbf{B}$ at a velocity which is independent of B, and which is much greater than v_y. The conductivity along $\mathbf{B}$ is therefore very high, giving rise to a current which cannot circulate vertically, for the lower ionosphere is not a conductor; the motions are thus mostly horizontal.

Given the system of winds and the conductivity, we arrive at systems similar to Figure 30, from which one can calculate the diurnal and seasonal variations of the terrestrial magnetic field. One can also follow the inverse procedure, and derive the system of winds.

In the F region, the velocity $\mathbf{u}$ of the neutral particles is negligible. However, if we take the mechanical forces into account, we can explain important effects in the motion of the ionized particles. Writing an explicit expression for the vertical component of the ambipolar diffusion, $v_h = -v_z \sin I$ where I is the magnetic inclination, we find

$$v_h = -\frac{g}{\nu_e}\left(1 + \frac{2H}{n}\frac{\partial n}{\partial h}\right)\sin^2 I,$$

where H is the scale height and n the number density. The velocity v_h increases with h, so that this diffusion becomes the dominant phenomenon above 300 km. One can explain the existence of an ionization maximum which is not directly controlled by the Sun by introducing v_h into the motion term M of the electron density equation.

CHAPTER IV

THE OUTER IONOSPHERE AND THE VAN ALLEN BELTS

1. General Considerations

For a long time our knowledge of the ionization of the atmosphere stopped at the level of the F-region maximum, for there were no means of studying it at greater heights. No one questioned the existence of charged particles at much greater altitudes, but it was difficult to know where the ionosphere ended.

The fact that electrons remain present as far out as several terrestrial radii became a certainty after the interpretation of the propagation of 'whistlers' given by Storey in 1953. We have already examined this phenomenon of dispersive wave guiding along the lines of force of the terrestrial magnetic field.

Other methods confirmed these conclusions. Among them, we note the use of the Faraday effect. When radio waves pass through a magnetized plasma, their plane of polarization is rotated. This technique was applied to waves reflected from the Moon, and appreciable electron densities were found up to thousands of kilometers from the ground.

Nevertheless, the information remained very fragmentary and it was, once again, only through experiments made *in situ* by rockets and satellites that an overall picture of the very high ionosphere was obtained.

It is not possible to dissociate the neutral part of the atmosphere from the ionized part, even though their distribution is very different.

The neutral particles are distributed according to the laws of hydrostatics, assuming that collisions are frequent and that the mean free path is very small. But the mean free path is already 1 km at an altitude of 250 km, and a little higher up it becomes equal to the scale height H. From this level onward, we must consider the motion of the individual particles, which are no longer maintained in the same position for lack of collisions. Each of them describes a ballistic trajectory in the gravitational field, a situation which defines the critical level of which we have already spoken. This is the base of the exosphere.

The distribution of the ions and electrons is governed by the same factors, but new influences are added and become dominant.

Let us mention three of these new factors:

– the electrically charged particles 'see' each other from much farther away than the neutral particles, for they attract or repel each other with the so-called coulomb force which decreases only as the square of the distance. We say that their effective cross-section for collisions is much greater, where the word 'collision' does not mean a simple impact but a large and steady modification of the trajectory;

– there can be no large-scale separation of the ions and the electrons, for this would produce an electric field which would draw them together again. The scale height for ions is twice as great as that for neutral particles of very similar mass;

– the guiding effect of the terrestrial magnetic field keeps the charged particles near the Earth.

Moreover, we must distinguish between thermalized and non-thermalized charged particles, depending on whether or not the density is significant. This consideration leads us to divide the upper atmosphere into two large regions, namely:

– the upper ionosphere, where the concentration of charged particles is great enough so that they affect each other;

– the Van Allen belts, discovered by the first satellites and studied in great detail since then. The magnetic field completely governs the motion of the ions and electrons, making them travel along the lines of force and causing the drift motions which we have already examined in Chapter II.

We shall consider in succession these three aspects of the very high atmosphere, collecting the first two in the same section.

2. The Exosphere and the Upper Ionosphere

A. NEUTRAL PARTICLES

The neutral particles dominate up to altitudes of several thousand kilometers. Different techniques are required to measure their concentration. Once this parameter is known as a function of altitude, the pressure, temperature, and molecular weight can be deduced.

One of the most effective methods of obtaining this density consists of following the motion of a satellite over a long period. In principle one can control the altitude and velocity of injection into orbit.* The orbit chosen must be a highly eccentric ellipse. The air encountered by the satellite introduces a force of friction which can be written

$$F = \frac{\rho}{2} K V^2 S ,$$

where ρ is the density we wish to determine, V the velocity of the satellite, S its 'effective' cross-section, and K a coefficient of friction characteristic of the medium.

This force is much greater near perigee than near apogee. Beyond 1000 km it can even be considered negligible, while other types of force of friction appear: solar radiation pressure, electrostatic effects, attraction of the Sun and Moon, and irregularities in the gravitational field.

The satellite is therefore braked primarily near its perigee, causing a progressive lowering of the apogee with each revolution, and thus a decrease in both the eccentricity (Figure 31) and the period. From the decrease in period, we deduce ρ near the

---

* Consider an elliptical trajectory around the Earth. At any point of this trajectory, the velocity of the satellite is well determined both in magnitude and direction. It is therefore sufficient to bring a spacecraft to this point and to give it the proper velocity, in order to cause it to follow the desired ellipse.

perigee. Moreover, the direction of the perigee as seen from the Earth is not immut-
able. The orbit turns slowly in its own plane, while the Earth turns around the Sun.
Thus the perigees pass from day to night, enabling us to bring out the role played by
the latitude, the longitude, and the temperature, if the lifetime of the satellite is long
enough.

The pressure can be deduced from the measurements of ρ, so long as the laws of
hydrostatics remain valid – that is, between 180 and a few hundred kilometers. Below
these altitudes, the satellite does not remain in orbit long enough. At higher altitudes,
there are too few collisions. The temperature is obtained by assuming a perfect gas. We
must know the molecular weight, which itself depends on the temperature as a consequence of the diffusion of light elements towards higher altitudes. But since the temperature does not vary greatly (Figure 2), this ambiguity is not very serious.

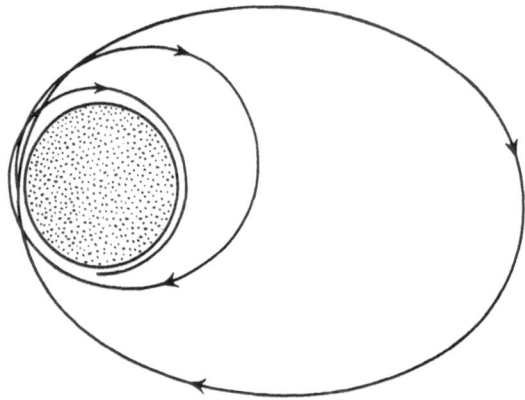

Fig. 31.   Distortion (very exaggerated) of the successive orbits of a
satellite as a result of braking in the atmosphere.

From all the measurements made since 1957, we conclude:
– that there is a large daily variation;
– that there is a clear correlation with solar activity;
– that there is a semiannual variation.
The density is maximum during the afternoon, and minimum in the second half of
the night. The effect increases with altitude: it ranges from a few percent at 200 km to
enormous relative values at 600 km. The temperature varies in the same way, but the
ratio between day and night is no greater than 3/2.

The correlation with solar activity was established in 1958. A periodicity of twenty-
seven days has been observed, as well as an increase in density during solar flares (cf.
Chapter VI, p. 79). Jacchia has shown that the temperature of the exosphere conforms remarkably well to the flux of energy received from the Sun at a wavelength of
10.7 cm.

Finally, the semiannual variation produces a temperature maximum at the time of
the equinoxes. The various illustrations in Chapter I show the variation of the atmospheric parameters with altitude.

We can attempt to give a theoretical basis for the concept of the critical level h_c (the base of the exosphere), considering for simplicity's sake only a single constituent. Let us follow a particle of radius r as it rises vertically. It will encounter all those particles whose centers approach it within a distance $2r$ – that is, the particles contained in a column of cross-section $4\pi r^2$. To find the number of particles in question, we must consider their distribution $n(h)$ above the critical level h_c which we wish to determine. If this distribution is hydrostatic, $n(h)$ is an exponential of the form $n_c \exp(h_c - h/H)$, which leads to the value

$$n_c = \frac{1}{4\pi r^2 H}$$

for the density at the base of the exosphere. Since r is on the order of an ångström unit and H is on the order of 100 km, we find several tens of millions of neutral particles per cubic centimeter. This value is appropriate for $h_c = 500$ or 600 km.

B. THERMALIZED CHARGED PARTICLES

Entirely different methods make it possible to obtain the constitution of the upper ionosphere. They have been developed only recently.

A priori, we know that the density of charged particles decreases continuously beyond the level of maximum ionization of the F region. The rate of decrease was overestimated for a long time; in reality it is much lower than that of the neutral particles, for the reasons set forth at the beginning of this chapter. As a result, the degree of ionization can increase with altitude until it reaches unity.

Since radio sounding from the ground is not suitable, other methods – direct or indirect – must be used to study these phenomena. We shall restrict ourselves to the three principal methods.

(a) We already know that Storey contributed the first proof of the existence of electrons at very high altitudes. Let us make a more detailed examination of his method. Since electromagnetic waves of very low frequency (kHz) are guided in their propagation by the lines of force, they provide us with a means of studying very high levels – the higher the magnetic latitude at which the transmitter is located, the higher the altitude that can be studied. The signals can be produced naturally by lightning discharges, which give a whole spectrum of frequencies that travel in the fashion described in Chapter II. This propagation is dispersive as a consequence of the existence of electrons in space. The wave remains for the longest time in the region where the induction B is lowest – that is, above the magnetic equator. Making an *a priori* assumption as to the form of the electron distribution function, we can find its absolute value by means of the signals $f(t)$ recorded as in Figure 14. We arrive at 10^2 or 10^3 electrons/cm^3 at around 15 or 20 000 km. A more exact treatment brings out the 'nose effect', which means that there exists a frequency for which the transit time is a minimum. This frequency gives us the minimum gyrofrequency along the line of force, and its altitude. Such measurements have given rise to many studies which bring out the annual, undecennial, and abnormal variations.

(b) A second method consists of receiving echoes of signals sent from the ground.

The frequency must be either much lower or much higher than the critical frequency of the F region.

In the first case, we are really creating artificial whistlers, with the advantage of being able to control them. This is an expensive procedure.

In the second case, we rely on two phenomena:

– the Faraday rotation mentioned above. The total rotation observed in a lunar echo is related to the total electron content of the region traversed. Thus we find that during the day, there are three times as many electrons above the peak of the F layer as below it. At night, the ratio is even greater;

– the incoherent scattering of a wave by electrons. Inhomogeneities in the ionization distribution give rise to backscattering. Since the amount of energy sent back is small, a powerful radar must be used. It is possible to receive echoes out to 5000 km.

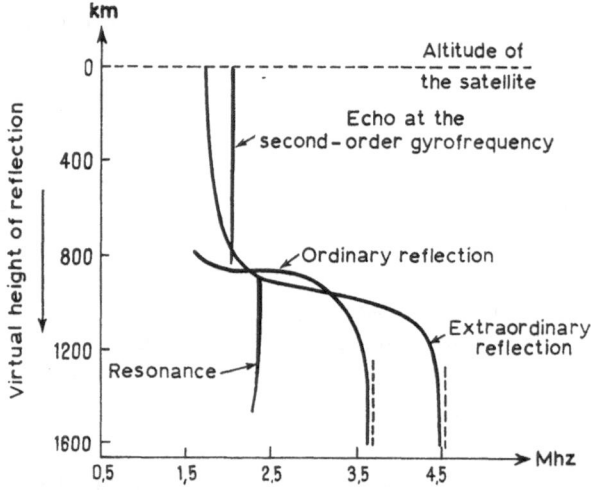

Fig. 32.   Ionogram recorded by the satellite *Alouette I* (after E. L. Hagg).

(c) Measurements *in situ* were first made in 1960, and really came into their own in 1962 with the launching of the satellite *Alouette* in an almost polar orbit about 1000 km from the ground. The altitude was chosen to be high enough to take in almost the entire thickness of the ionosphere, but low enough to retain a period of revolution that would be short compared to the time scale for changes in the ionosphere. Several satellites of the same type have been launched since then.

Figure 32 presents one of the ionograms recorded by *Alouette*. It is to be compared with those obtained from the ground and shown in Figure 24. We recognize the ordinary and extraordinary curves, which enable us to obtain the ionization profile of the upper slope of the ionosphere, once we have made the transformation to real heights. Echoes of a new type appear, particularly on multiples of the gyrofrequency. This phenomenon is not observable from the ground, where the absorption at these frequencies is much too high because of the presence of the D region. Other oblique reflections are also possible, on account of a phenomenon comparable to whistlers.

We may wonder whether the electron distribution obtained in this way is the same as the ion distribution. We know that a separation of these two types of charged particles gives rise to an opposing electric field. When we take this field into account, we obtain a scale height for the ions,

$$H_i = \frac{k\,(T_i + T_e)}{m_i g},$$

which is twice that for the neutral particles if the temperatures of the ions and the electrons are assumed to be equal and constant.

In reality, there are several types of ions, a fact which makes the problem much more complicated. The distribution of the less abundant ions will be very different from that of the more abundant ions, for the former will be subject to the electric field of the latter. The result is a tendency to find the lightest ions in greater and greater quantities as one goes away from the Earth. The ion O^+ dominates in the central part of the F region; but it is succeeded by helium nuclei, and finally protons become the only remaining constituent. They form the *protonosphere*, where collisions are finally negligible. We have then entered the Van Allen belts.

3. The Van Allen Belts

A. DISCOVERY OF THE BELTS

The permanent existence of charged particles confined in the terrestrial magnetic field above the ionosphere was discovered at the beginning of 1958, immediately after the first flights of the American satellites *Explorer I* and *III*. The phenomenon was later confirmed by *Sputnik III*. This discovery was a surprise, although it could have been predicted as a consequence of many previous studies.

J. A. Van Allen has given his name to the whole portion of space where the charged particles are primarily subject to the terrestrial magnetic field, since collisions occur only as very rare accidents. The discovery originated in a desire to make a better study of the cosmic rays, or very energetic particles, that come to us from beyond the Earth. For this purpose, Van Allen and his colleagues at the University of Iowa placed on board the first satellites Geiger counters which were sufficiently well plated to avoid any foreseeable saturation during the counting. However, the recorded counting rates proved to be much higher than expected. Van Allen proved that this was caused by charged particles bound to the Earth, for the lines of equal flux coincided rather well with the lines of force of the terrestrial magnetic field.

A few months later, in July 1958, a new satellite, *Explorer IV*, undertook a more detailed study; it investigated the space distribution, the energy, and the change in the electron density with time. In this way, the volume of the trapped particles was determined. In addition to the fact that the lines of equal flux were close to the lines of force, it was quickly observed that the counting rate was smallest along the lines of force that touch the globe at a geomagnetic latitude of around 50°. The existence of this minimum led to the definition of two separate belts, the inner belt and the outer belt.

The counters measure only an energy flux, from which the derivation of the spectrum is a delicate operation; but still there was every reason to suppose that protons and electrons were present. Using the first *Explorers* and *Pioneer I* and *III*, Van Allen

was able to draw the outline in Figure 33, where all the observations have been reduced to the same magnetic meridian.

The existence of two belts is really apparent only for particles which penetrate plates
of density 1 g/cm³; the distinction is connected with the nature of the particles: protons in the inner belt, and electrons in the outer belt. For convenience, we shall continue to make this distinction, even though charges of opposite sign exist in both belts.

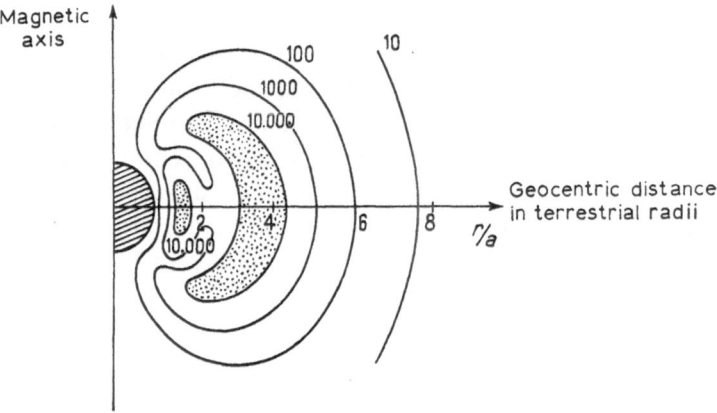

Fig. 33.   First counting rates obtained by the American satellites, reduced to a single
magnetic meridian (Van Allen and co-workers).

B. THE INNER BELT

The particles detected have been identified by means of nuclear plates on which they
leave traces during rocket flights. When these plates are recovered and developed, they
yield the nature and the energy of the particles. Figure 34 shows the proton spectrum

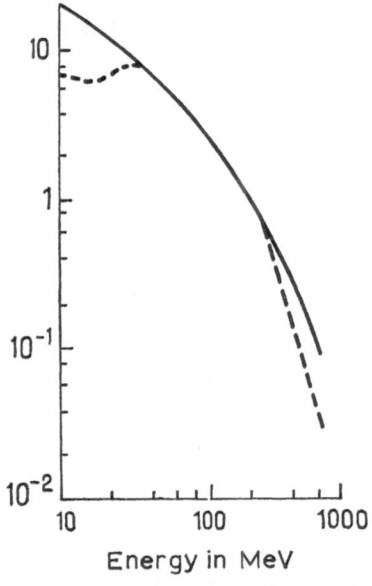

Fig. 34.   Proton spectrum in the inner belt.

determined from several flights. We see that it extends from 1 to 700 MeV and that it becomes flat below 40 MeV, where a slight irregularity is observed. Figure 35 indicates the flux distribution in a more elaborate fashion.

The origin of these charged particles is a problem that has not yet been clarified. Since the protons do not originate in a local neutral gas, they must be injected from outside. *A priori*, such an operation is forbidden by the presence of the magnetic field. There must be, however, one or more mechanisms that remove this theoretical prohibition, which is also contradicted by the existence of disturbances (Chapter VI).

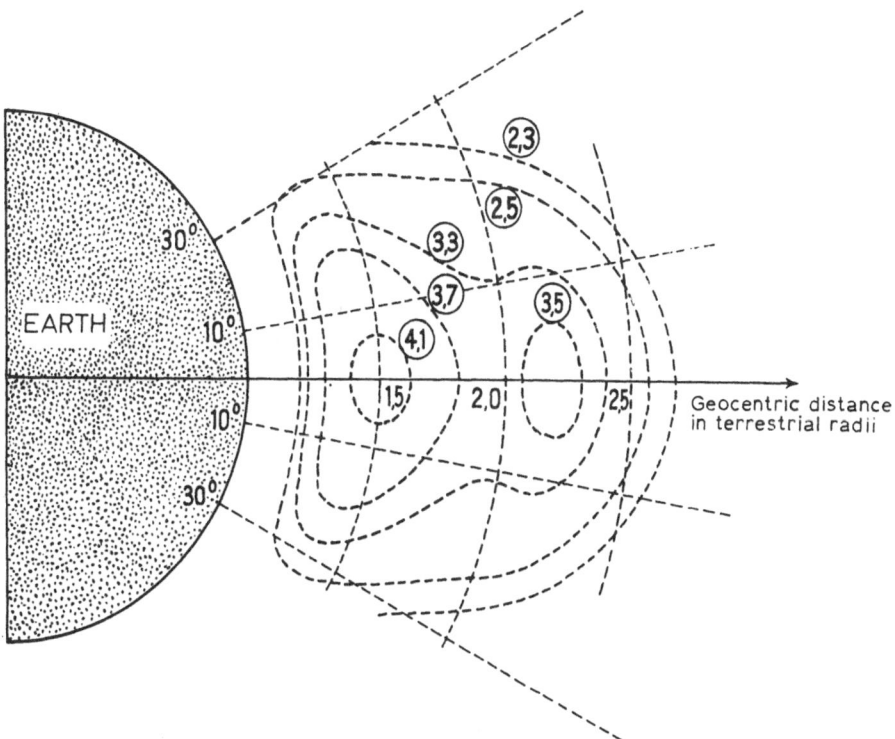

Fig. 35.   Isoflux contours for protons with energies between 40 and 110 MeV in the inner Van Allen belt (McIllwain, 1963).

Most specialists now agree that the particles are of cosmic origin. Two principal mechanisms have been and are being studied; we shall consider them in turn.

Protons constitute the principal component of the cosmic rays which continually bombard the Earth. They encounter the nitrogen and oxygen nuclei that form our atmosphere. The resulting collisions give rise to neutrons. Since the neutrons have no electric charge, they are not sensitive to the terrestrial magnetic field and can escape freely. As they are only slightly absorbed by air, some of them will rise to very great heights after being emitted in the dense layers. But the neutron is unstable outside the atomic nucleus. It is transformed into a proton and an electron after a time whose mean value is 12 min, according to the reaction

$$n \rightarrow p^+ + e^- + \nu + 780 \text{ keV}.$$

Most of the energy thus liberated is divided between the electron e^- and the neutrino ν, while the proton retains in practice the energy of the parent neutron. If this proton is born with a velocity suitable in magnitude and direction, it is injected into the terrestrial magnetic field without having to cross Störmer's forbidden regions, thus continually replenishing the inner belt.

An objection has been raised to this model: considering the cosmic-ray flux received by our planet, the protons would have to survive for at least a hundred years in order to account for the observed densities. This is an enormous time, for the protons are threatened by the collisions they may undergo with electrons at any point, and with neutral particles near the mirror points. The stability of the belt would thus appear to contradict the theory that it originates entirely in neutrons.

Other injection processes are possible, although less obvious than the preceding one; the principal such process is the acceleration of particles by the terrestrial magnetic field. The energy of an electrically charged particle in any static field is invariant, as we have seen. Therefore such an acceleration must be due to a temporal variation of the magnetic field, bringing about a true adiabatic compression. Since the field is anisotropic, two types of compression are conceivable:

– The first is called longitudinal, and is known as Fermi compression. Let there be a particle circulating on a tube of force between two mirror points. An increase in the induction brings these points closer together, and the longitudinal velocity of the charged particle is increased with each reflection, as for any projectile striking a moving surface.

The energy of the charged particle varies as the inverse square of the length of the tube of force.

– The second, called transverse compression, is nothing but the betatron acceleration of the nuclear physicists. If the induction increases slowly, the adiabatic invariant μ is conserved, and the transverse energy $W_\perp$ of the charged particle increases because

$$|\mu| = \frac{\frac{1}{2}mv_\perp^2}{B} = \frac{W_\perp}{B}.$$

Now the radius of gyration R_g varies as $v_\perp/B$, so that the cross-section of the tube of force to which the charge belongs, which varies as $v_\perp^2/B^2 \sim W_\perp/B^2$, decreases. There is indeed compression, but this time with two degrees of freedom.

Finally, let us point out that protons of energy less than 30 MeV have been detected near the outer edge of the belt, at about 4500 km of altitude. They may originate in neutrons resulting from the arrival of solar protons in the atmosphere near the poles (cf. Chapter VI). Since they are of lower energy, they are restricted to high altitudes. More recently, a new maximum in the radial distribution has been discovered, at an altitude of around 7500 km near the magnetic equator. Its origin is uncertain.

If the creation mechanisms remain poorly understood, the situation is not much better as regards the loss mechanisms. Among them, we may mention:

– Gradual losses due to collisions with the components of the atmosphere. These collisions modify the direction of the velocity, causing the particle to enter the 'loss cone'. *

– Charge exchanges, which lead to the same result.

– The existence of hydromagnetic waves, causing variations in the induction. The particle ceases to be trapped if its radius of gyration becomes comparable to the scale of the inhomogeneity.

Little is known about the electrons of the inner belt. The measured energy spectrum is in good agreement with that deduced from the neutron theory, at least above 400 keV. There is indeed a cutoff at 780 keV, which is an argument in favor of this hypothesis. On the other hand, the low-energy electrons (there are many of less than 100 keV) must have a different origin. Thus there is still an unknown source to be found.

C. THE OUTER BELT

One goes gradually from the inner belt to the outer belt, a region where electrons are detected in large quantities. Its shape follows that of the lines of force. The heart of this belt is situated above the equator at about twenty thousand kilometers from the center of the Earth, and it is bounded in the direction of outer space by a sharp limit which is also the limit of the terrestrial magnetic field (cf. Chapter V).

The first flux measurements made at large distances from the Earth gave impressive values, corresponding to counting rates greater than 100 000 counts/s.

Explorer VI, launched in 1959, was the first satellite that had a high enough apogee (42 000 km) to pass several times through this region. It carried coincidence counters, ionization chambers, and scintillation counters – but the three types of measurement did not yield the same space distribution for the charged particles.

Explorer XII (1961) cleared up the confusion; orbiting between 83 000 and 6700 km from the center of the Earth, it provided measurements over a period of three months. Its detection instruments were very diversified, in order to be able to study a spectrum which might range from several eV to 1000 MeV.

It was then realized that, contrary to what had previously been thought, the energy of the charged particles was large enough to pierce the armor of the instruments used before. The X-rays resulting from the interaction of electrons on the surface were not the only ones counted, and the 100 000 counts/s observed at the time were not due to *these* X-rays. The flux which had been previously estimated at 10^{11} electrons/cm^2/s was really no greater than 10^6, for the efficiency of the electron counting had been greatly underestimated.

Since that time, many measurements have been made and correctly interpreted, but they are still not enough to give a detailed description of the belt. The energy spectrum is rather flat in the heart of the belt, and most of the flux comes from electrons between 100 and 1000 keV. The flux becomes softer at greater distances from the Earth. Near the boundary, the electrons have energies between 0.2 and 20 keV.

---

* The loss cone is a geometrical translation of the existence of mirror points. Let us consider a particular point M. Only those charges whose velocities form an angle θ greater than a critical value θ_c with the line of force passing through M will come to a mirror point in the lower atmosphere. This results from the considerations in Chapter II.

The outer belt must be thought of as an extremely tenuous medium. The total mass of the particles that form it is scarcely greater than 15 kg.

As for the protons that accompany the electrons, they have been little studied as yet. They are found in large numbers at about three or four terrestrial radii, with a flux of 10^8 for energies of several hundred kilo-electron volts. These energies are much lower than those of the protons in the inner belt. On the other hand, the flux is much too large to be explained by the neutron hypothesis.

The outer belt presents a number of question marks. Although the protons studied by several satellites (particularly *Explorer XII, XIV*, and *XV*) have been shown to be remarkably stable in flux and energy, this is not the case with the electrons, which undergo large temporal variations in number, energy, and flux – especially at the time of the magnetic storms characterized by variations in the terrestrial magnetic field and the arrival of charged particles from the Sun. The flux at low energies is then observed to increase, while the high-energy electrons are lost. After the storm, they reappear in the same numbers, and the original structure is recovered. Variations of a factor of 100 have been observed in a few hours.

This solar effect is still very poorly understood. We shall come back to it in the following chapters.

D. ARTIFICIAL BELTS

Since the time of the first space studies, seven nuclear explosions have been produced at high altitudes.

The first, known as the Argus experiments, were carried out in 1959 to study the trapping of charged particles in the terrestrial magnetic field. In 1959, Christofolos suggested liberating a large number of electrons in this way, and following their evolution by means of satellites. In fact, three charges, each equivalent to one kiloton of T.N.T.,* were detonated at an altitude of 480 km above the Atlantic. They produced artificial belts that were followed by the satellite *Explorer IV*, which had been launched for this purpose. The electrons diffused rapidly around the Earth, forming a layer which was maintained for about a week without much radial motion.

With the Starfish experiment (1962), the Americans changed scale. A charge of 1.4 megatons was exploded at the same height above the Pacific. A very large electron belt resulted. It was maintained much longer than expected, disturbing the natural state of the Van Allen belts for several years. Many satellites have had time to study its electron spectrum and space distribution, which reached an altitude of more than 3000 km. We do not understand how electrons originating in radioactive decays and fission products could have been diffused so high, nor why the observed spectra depend strongly on altitude.

At the end of 1962, three other artificial belts were created by the Russians, more as nuclear tests than as space studies. We do not know the experimental conditions, but it appears that the explosions took place much higher up, perhaps above 4000 km. Be that as it may, the belts decayed very rapidly and scarcely polluted the natural belts at all.

* One kiloton is about 5×10^{12} J.

THE BORDERS OF THE TERRESTRIAL ENVIRONMENT

1. Introduction

It was more than thirty years ago that Chapman and Ferraro came to the conclusion that, in addition to its light, the Sun emitted charged particles that arrived in the neighborhood of the Earth. This flux, intended at the time to explain magnetic storms, was considered to be a transitory phenomenon accompanying a *solar flare* – a term designating a bright light which suddenly appears on the surface of the Sun and is followed by magnetic and ionospheric disturbances on our planet.

For his part, Bierman pointed out in 1951 that the comet tails which always point in the opposite direction from the Sun were subject not only to radiation pressure, but also – and, indeed, primarily – to the pressure of a permanent plasma accompanied by a magnetic field.

Studying the solar corona, Parker gave the first theoretical basis for this phenomenon and coined the term *solar wind*.

All these ideas were confirmed, beginning in 1958, by the artificial satellites and

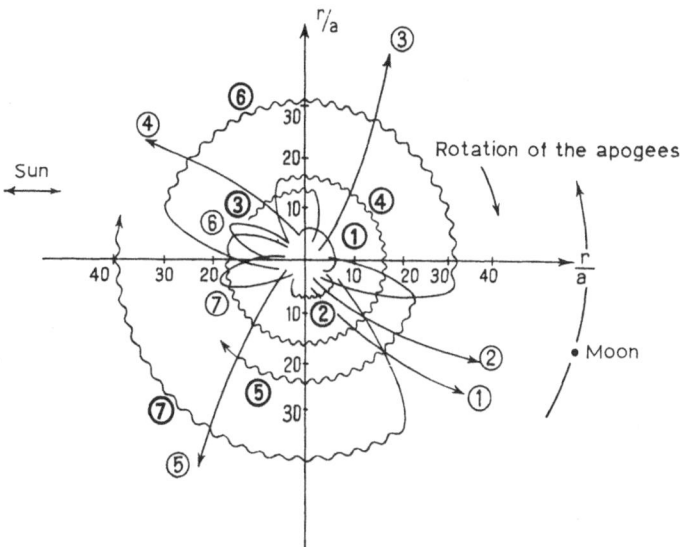

Fig. 36.   Regions explored by various satellites and space probes.
(1) *Electron II*; **(2)** *Explorer VI*; **(3)** *Explorer XII*; **(4)** *Explorer XIV*; **(5)** *OGO-A*; **(6)** *Explorer XVIII (IMP I)*; **(7)** *Explorer XXVIII (IMP 3)*.
(1) *Pioneer VI*; (2) *Lunik II*; (3) *Mariner IV*; (4) *Pioneer IV*;
(5) *Pioneer V*; (6) *Pioneer III*; (7) *Pioneer I*.

space probes which have provided us with the means of studying the terrestrial magnetic field and the characteristics of the plasma in the regions they traverse. From this point of view, the most interesting vehicles are the satellites having very eccentric orbits with apogees of several tens or even hundreds of thousands of kilometers. The table gives a list – already a long one – of the vehicles which have provided the most information in this field. We note that their lifetime and their apogees have continually been improved. This is due to progress in rockets and in the electronic equipment and electrical energy supplies on board. With a long useful lifetime, these satellites can sweep out a considerable volume of space, principally on account of the phenomenon of precession of the apogees.

TABLE I

Vehicle	Date	Inclination	Lifetime (days)	Distance (terrestrial radii)
Pioneer I	11-10-58		1	3.7 to 7.0
Explorer V	7- 8-59	47°	61	2 to 7.5
Pioneer V	11- 3-60	Solar orbit	50	5 to 9.0
Explorer X	25- 3-61	33°	2.2	1.8 to 7.0
Explorer XII	16- 8-61	33°	112	4 to 13.5
Explorer XIV	3-10-62	33°	300	5 to 16.5
Explorer XVIII	27-11-63	33°	181	<32
Electron II	30- 1-64	61°	90	3 to 11.6
OGO A	5- 9-64	31°	> 500	3.8 to 24.3
Explorer XXI	4-10-64	34°	150	6 to 15.9
Explorer XXVIII	29- 5-65	33°	> 250	<42
Pioneer VI	16-12-65	Solar orbit between Earth and Venus		

The most direct consequence of the existence of a solar wind is that the terrestrial magnetic field is confined to the inside of a cavity called the geomagnetic cavity or *magnetosphere*. Calculations and measurements agree in fixing its surface at a distance of about ten terrestrial radii on the sunlit side. The boundary surface has been given the name of *magnetopause*. Measurements on the night side are less numerous, and it is more difficult to derive a general picture from them. They indicate a very different structure, which produces a *magnetic tail* extending much farther from the Earth. The magnetosphere turns around the Earth, following the position of the Sun.

We can thus distinguish three regions:

(a) The interplanetary space which is not disturbed by the presence of the Earth and its magnetic field;

(b) A region called the transition zone or field-plasma interaction zone;

(c) The magnetosphere, which contains the terrestrial magnetic field with the Earth at its center.

A new surface has been discovered. It is located beyond the magnetosphere. We shall call it the *shock front*.

2. The Solar Wind and the Interplanetary Magnetic Field

The charged particles emitted by the Sun must come from its equatorial region if they are eventually to affect the Earth, for the axis of solar rotation makes an angle of only

$7° 3'$ with the normal to the plane of the ecliptic. If we assume that the particles are ejected vertically, as seems to be the case, we must take the rotation of the Sun into account in order to obtain the real initial velocity. The velocity of ejection varies with the degree of solar activity. In a so-called quiet period, it seems that the velocities of the charged particles in the vicinity of the Earth are between 300 and 500 km/s. During their voyage, these particles are attracted by the gravitation of the Sun, but repelled by its radiation pressure. They are accelerated as they go farther from the Sun, and finally attain a limiting velocity much greater than that necessary to escape from the Sun's attraction.

When the emission at the surface of the Sun is modified in space or time, the Earth may pass through a different solar wind a short time later. The Earth intersects the trajectories of the individual particles at an angle which depends upon the characteristics of the emission. To be specific, let us say that it takes about a day for the plasma to reach the Earth.

Space probes have shown that the solar wind is composed of electrons, of protons, and to a lesser degree of helium nuclei – as we might have predicted, knowing the chemical composition of the Sun. The ejected plasma is electrically neutral. Whatever the sign of their charge, the particles travel at the same velocity, so that the net electric current is zero.* The particle density is on the order of several protons per cubic centimeter at the distance of the Earth. It seems to decrease as $1/r^2$ (where r is the distance to the Sun). The resulting mass loss from the Sun, assuming the emission to be isotropic, is only one part in 10^5 per billion years. This factor represents only one-tenth of the losses due to thermonuclear reactions. Initially, only the region of space near the Earth – that is, 150 million kilometers from the Sun and near the ecliptic – was studied, except for the results of the Mariners and Venusiks. But the American satellite *Pioneer VI*, launched in an orbit around the Sun at the end of 1965, passed within 123 million kilometers of the Sun. Its observations have clarified many points concerning the physics of space.

The space between the Earth and the Sun is the site of a weak magnetic field on the order of 5γ ($1\gamma = 10^{-5}$ G), which plays an important role, for magnetohydrodynamics shows that the behavior of a plasma depends on whether its kinetic energy per unit volume (ρv^2) is greater than or less than the magnetic energy density of the medium through which it moves. In the first case, it is the plasma that dominates the field. If the field was already present, it is trapped by the plasma and carried along with it without being able to escape. In the second case, it is the field that guides the charged particles, as we have seen in Chapter II.

3. The Magnetopause

The Earth presents a unique obstacle to the solar wind, for it is endowed with its own magnetic field. In the case of the solar wind, we find ourselves confronted with a good conductor (the plasma) which approaches a magnetized dipole (the Earth).

---

* A current is composed of electrons moving in one direction and ions moving in the opposite direction. In metals, the ions are motionless. In electrolytes, both types of charge take part in the current. In the present case, the two currents cancel each other.

The interaction problem is complicated, for the gas is a three-dimensional, deform-able conductor. The problem of a plane, infinite conductor moving with respect to a dipole was treated by Maxwell himself in another connection. His solution gives a good idea of the deformation to which the lines of force are subject (Figure 37). We see that a certain asymmetry arises. The currents induced in the conductor give rise to an induction which, according to Lenz's law, produces a complete screening effect to-wards the outside – that is, on the side opposite the dipole. On the side nearer the di-pole, the induction is increased, as if an image-dipole were present. The resulting induction does not have a component perpendicular to the conductor as it normally would, and the lines of force are tangent to the conductor except at two points, the neutral points Q and Q' of Figure 37. There, the induction is zero. The existence of

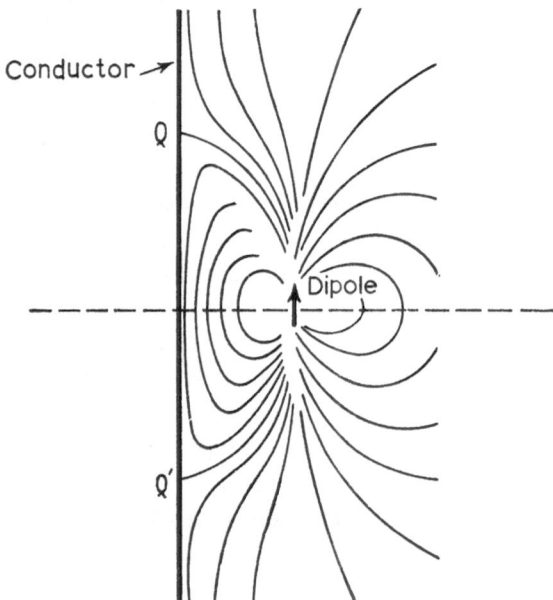

Fig. 37.   Effect of a conducting plane on a magnetic dipole. Distortion of the lines of force.

such points (or, more exactly, of such lines) in space can play an important part in en-abling charged particles to penetrate the magnetosphere, thus explaining at least in part such phenomena as the magnetic storms of which we shall speak in the following chapter.

The induced currents give rise to electromagnetic forces which act in a non-homo-geneous fashion on the plane, tending to deform it. But since the magnetopause is not a plane, nor even a rigid surface, it can adapt itself to this situation.

One of the basic problems consists of determining the shape of the magnetopause. Much theoretical progress has recently been made, starting from an outline similar to the above. Several authors have obtained numerical results by assuming a field-free incident plasma, a zero temperature, and an angle of incidence normal to the terres-trial dipole.

The surface obtained is essentially hemispherical on the day side, as is indicated in

Figure 38, which represents a meridian section containing the Sun. Two neutral points can be observed, at magnetic latitudes of about 70°. On the night side, the magnetopause becomes essentially cylindrical, with a diameter of 30 to 40 terrestrial radii.

When the wind is not blowing in the plane of the magnetic equator, which is generally the case except at the solstices, the shape of the magnetopause is slightly modified. Primarily, it undergoes a wobbling effect which depends on the value of the angle λ.

The first observation of this boundary was made by *Explorer X* in 1961. The relatively high magnetic induction (20 to 30 γ) which the satellite encountered inside the magnetosphere changed rapidly in direction and amplitude, while the plasma sounders indicated a greatly increased signal.

These observations were confirmed by *Explorer XII* in the same year, at a distance

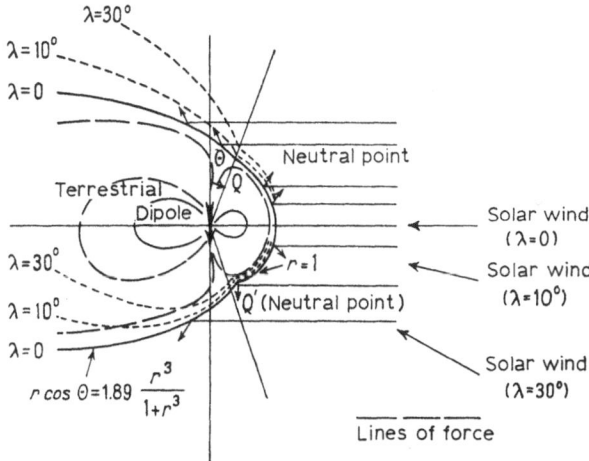

Fig. 38.   Effect of the solar wind (a conductor) on the terrestrial magnetic dipole.

between 8 and 10 terrestrial radii, and then by *Explorer XIV*, which orbited even farther from the Earth. However, it was the satellite *IMP 1* or *Explorer XVIII*, in 1964, which gave us the first overall view of the confined terrestrial magnetic field and of the transition zone (Figure 39).

Inside the magnetosphere, the induction has the predicted direction. Its intensity increases abruptly by a factor of 2 in the vicinity of the magnetopause, as indicated by the screening effect. In the first approximation, it is possible to estimate the distance of the magnetopause by writing the equation of equilibrium between the magnetic pressure and the kinetic pressure of the solar plasma, that is:

$$\frac{B^2}{2\mu_0} = mnv^2 \, ,$$

where m is the mass of the proton. Since $B = 2B_0 \, (a/r)^3$, where B_0 is the induction at ground level and at the equator, we find that the distance r of the magnetopause varies from 8 to 11 terrestrial radii when the proton density n goes from 2 to 10 protons/cm^3 (with $v = 400$ km/s).

The magnetopause (Σ) is not an ideal surface. It has a certain thickness and a structure. Near the equator, the charged particles arrive on a trajectory almost perpendicular to the lines of force. The ions, which have a momentum 2000 times greater than that of the electrons (their velocities being the same), penetrate more deeply into the terrestrial magnetic field. As a result, there is a tendency for the charged particles of opposite sign to separate, giving rise to an electric field which opposes this separation. This field is normal to (Σ) and directed towards outer space. The combination of this

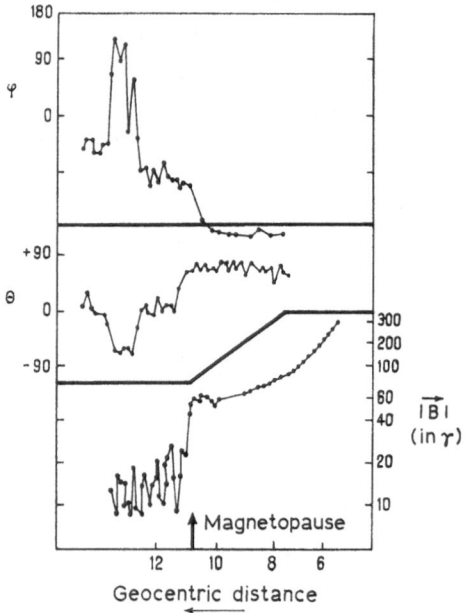

Fig. 39. Magnetic induction measured by *IMP 1* as it crossed the magnetopause. Note that |**B**| is double-valued at a geocentric distance of 10.8 a, and that there are discontinuities in the direction of **B**. (The dotted curves represent the theoretical values.)

field and the terrestrial magnetic field forces the charged particles to execute complicated drift motions, which bring about a very rapid (exponential) decrease in the induction. Calculations show that the magnetopause is very thin, a few tens of kilometers thick at the most. Satellites take only a few tens of seconds to cross it.

4. The Shock Front

The satellite *IMP 1*, which we have already mentioned, was the first to go completely beyond the magnetopause; after crossing a thick turbulent zone, it came to a region where the magnetic field became homogeneous and directed along a line joining the Earth and the Sun. As this turbulent zone could not be the same as the magnetopause – which, as we have seen, is thin – it was concluded that *IMP 1* had observed a second boundary which was identified as a 'collisionless shock front', and which had been predicted by several authors.

This shock front is interpreted by analogy with supersonic gas dynamics. The nature of the flow of a fluid around an obstacle depends on whether the velocity is subsonic or supersonic. In the latter case, a discontinuity in the physical properties develops, and is marked by a surface in front of the object. In the case of the Earth, it is not the planet itself but its magnetosphere that constitutes the obstacle. We are not dealing with the flow of a neutral gas, but with that of a magnetized and rarefied plasma. It can be shown that the frequency of collisions between the components of this plasma is negligible in comparison with the other frequencies which enter the problem, namely the plasma frequency and the gyrofrequencies of the ions and the electrons. On the other hand, the perturbations propagated in a plasma are not simple motions of

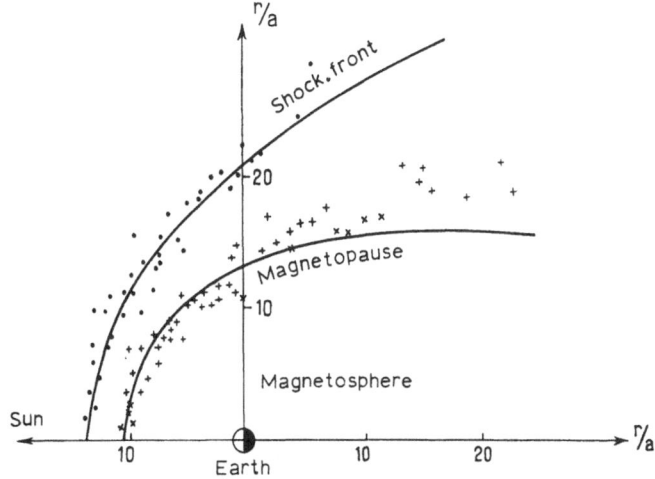

Fig. 40.   The magnetopause and the shock front, sunlit side.

neutral particles, as is the case with sound waves in a neutral gas. The speed of sound must be replaced with that of the so-called Alfvèn waves,

$$v_A = \frac{B}{\sqrt{\mu_0 \rho}},$$

where B is the interplanetary induction (cf. Chapter II).

With the values of the parameters measured by satellites and already cited, we find v_A between 25 and 90 km/s, always much less than the group velocity of the solar wind. Thus the flow is of the supersonic variety.

Figure 40 indicates the theoretical profile, and the observations of *IMP 1*. The agreement is good. If the parameters have been determined precisely, it is possible to invert the problem and to re-determine the velocity and density of the plasma.

Between the shock front and the magnetopause, the medium is turbulent. The magnetic field undergoes disordered variations, while the plasma flux ceases to be unidirectional and is said to be thermalized. The magnetic field sometimes becomes stabilized, taking on either the geomagnetic or the interplanetary form.

5. The Earth's Magnetic Tail

Although the lines of force are compressed on the sunward side, this is no longer true behind the Earth. The first satellites that traversed this region at large distances did not observe a magnetopause, in agreement with an outline such as that given in Figure 38. However, it must be pointed out that the theoretical determinations on which this profile is based assume that the interplanetary magnetic induction and the temperature of the plasma are zero. When these two factors are taken into account, the shape on the night side is greatly modified.

As early as 1960, Piddington, who was discussing magnetic storms, predicted that the lines of force originating above a certain latitude would not be closed on the night side. The satellite *IMP 1* confirmed this point of view. These lines of force form a sort of magnetic train which goes out to at least 30 terrestrial radii – half the distance from the Earth to the Moon. The magnetopause that maintains this train seems to have a radius of 20 terrestrial radii at large distances from our planet, contrary to the calculations which take into account the extra-terrestrial magnetic field and the non-zero temperature of the plasma.

IMP 1 was also able to measure the induction in a region located in the direction opposite the Sun and practically in the midnight meridian plane. This induction, which reaches 10γ to 30γ, is several times larger than the theoretical value obtained from an extrapolation of ground-level measurements. It is directed towards the Sun or away from it, depending on the location. A more complete reduction of the measurements indicates the existence of a layer free of magnetic field, which is sometimes traversed several times in a single revolution and which seems to have a thickness of 600 km.

If the axis of the terrestrial dipole were perpendicular to the plane of the ecliptic, this layer would be in the ecliptic plane. The differing obliquities of the lines of geographic and of magnetic poles, coupled with the rotation of the Earth, lead to a complicated oscillation of this layer. This explains why a 'slow' satellite can cross it several times.

The neutral layer is a permanent phenomenon which separates regions of equal but opposite induction. Figure 41 shows a strong day–night asymmetry, which is now considered to play a fundamental role in the Van Allen belts, in storm phenomena, and in the distribution of polar aurorae. We know practically nothing as yet concerning the regions located at 90° from the Earth–Sun line – in particular, nothing is known concerning the existence of neutral points.

Pressure equilibrium requires that this field-free layer should contain charged particles. It is reasonable to suppose that the fluxes measured by *Explorer XIV* and by the Russian satellite *Lunik I* on the night side, at large distances from the Earth, represent this phenomenon and not the existence of a third belt, as was thought at the time. The neutral layer appears to be a source of very energetic particles, making it possible to inject charged particles into the lower atmosphere, either into the belts or into the ionosphere. This layer is now considered indispensable for an explanation of how these regions are filled, and also of the mechanisms of aurorae. The plasma which is temporarily located in the neutral layer is thermalized, and directed towards lower altitudes

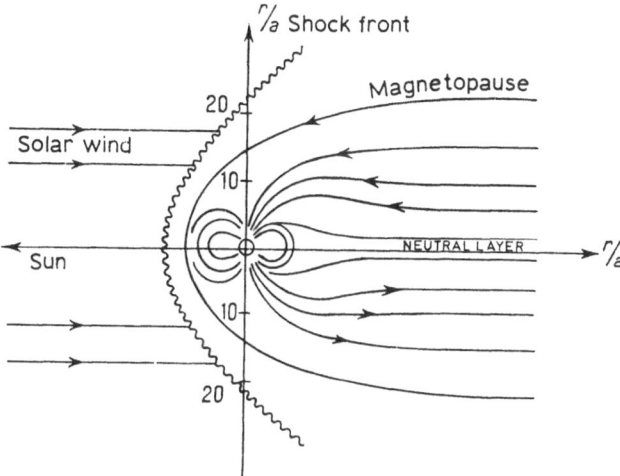

Fig. 41.   The Earth's magnetic tail and the neutral layer.

along high-latitude lines of force. The acceleration of the charged particles is believed
to originate in a loss of magnetic energy to the plasma, as occurs in certain laboratory
experiments.

Thus, according to one American physicist, the Earth in space looks like a comet:
– the Earth itself is the nucleus;
– the lines of force guide the charged particles all around the Earth with drift motions. As a whole, they correspond to the coma. Finally, the tail is represented by the
magnetic train.

This tail, just like those of comets, seems to result from the interaction of the solar
wind with two types of celestial body. In the case of a comet, there is no magnetic field.
The coma is the result of the evaporation of material from the surface. In the case of
the Earth, there are belts due to the trapping of charged particles in a magnetic field.
The existence of a magnetosphere is observed while, in the case of a comet, the coma
provides a plasma that captures the interplanetary field.

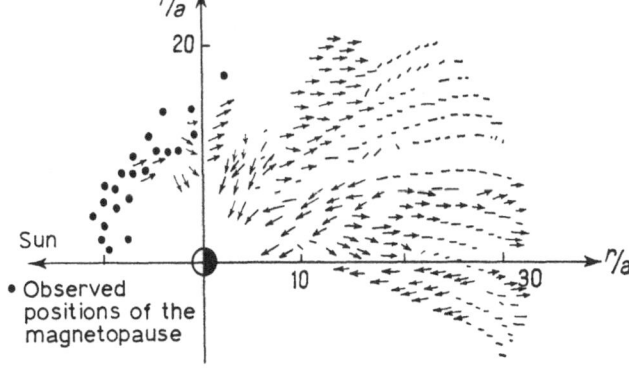

Fig. 42.   Measurements of the magnetic induction made by *IMP 1*, night side.

When we look towards other planets, we can predict that they must also leave behind a wake in the solar wind. This wake may not be the same as that of the Earth, for the Earth has its own magnetic field. Mercury and Venus do not seem to have magnetic fields, but there is every reason to believe that Jupiter does have one, for its surprising radio emission can be explained only by assuming a very eccentric dipole induction. Finally, the Moon has the singular property of crossing the Earth's magnetic tail at certain times in the lunar month. It must make its influence felt, and satellites such as *Luna X* which orbit around the Moon should make it possible to study this phenomenon.

CHAPTER VI

DISTURBANCES IN THE ATMOSPHERE

The electromagnetic and corpuscular emission of the Sun, such as we have described them in the preceding chapters, characterize the normal Sun with its cyclical variations. It frequently happens, especially outside the periods of minimum solar activity, that this emission is temporarily greatly increased. These increases are followed, after a greater or lesser delay, by disturbances in the upper atmosphere and in the terrestrial magnetic field, whose effects are noticeable even at ground level.

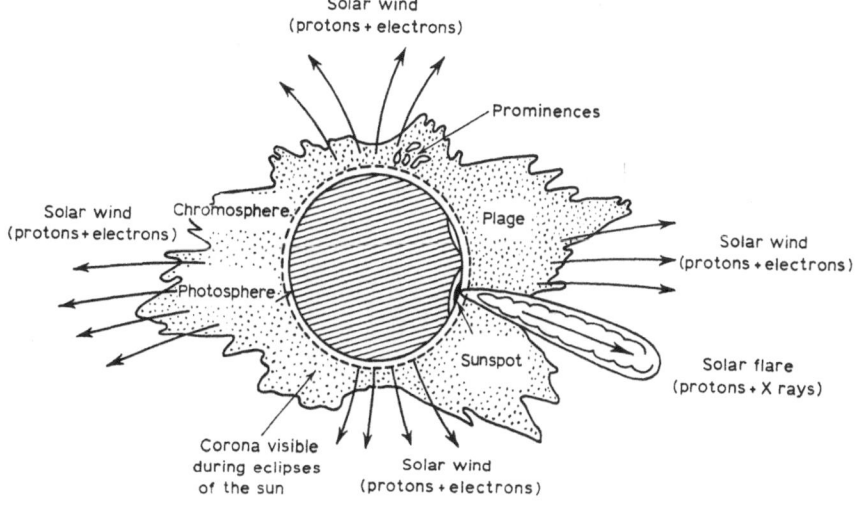

Fig. 43.   The Sun and its activity.

After having examined the nature of these solar anomalies, we shall describe their consequences for the Earth (Figure 43). Some of them affect the Earth as a whole, while others are restricted to high latitudes or to the sunlit hemisphere. The role of the terrestrial magnetic field appears to be essential where charged particles are concerned although we do not yet have a definitive explanation of its action. Disturbances caused by the Sun appear principally as:

– sudden ionospheric disturbances of short duration, principally involving the D region of the ionosphere;

– strong absorption of electromagnetic waves in the polar regions (polar cap absorption);

– magnetic storms or abnormal variations in the terrestrial magnetic field over the whole planet, accompanied by storms in all regions of the ionosphere;

– aurorae visible above high-latitude regions.

At the present time there is no satisfactory theory for these various phenomena, for solar–terrestrial relations are very complex and are still poorly known, despite recent progress which should continue rapidly. All these disturbances, which are important because of their effects on radio communications, can have repercussions on the conditions of space flights. Research is under way with a view to predicting these phenomena.

We shall examine the different effects in turn.

1. Electromagnetic Emission from the Active Sun and Sudden Ionospheric Disturbances

Sudden ionospheric disturbances are abrupt increases in the electron density at the level of the D region, which can result either in enhancing this region or in diminishing it. They have long been known to be connected with solar flares, which also have other consequences to be studied in the following paragraphs.

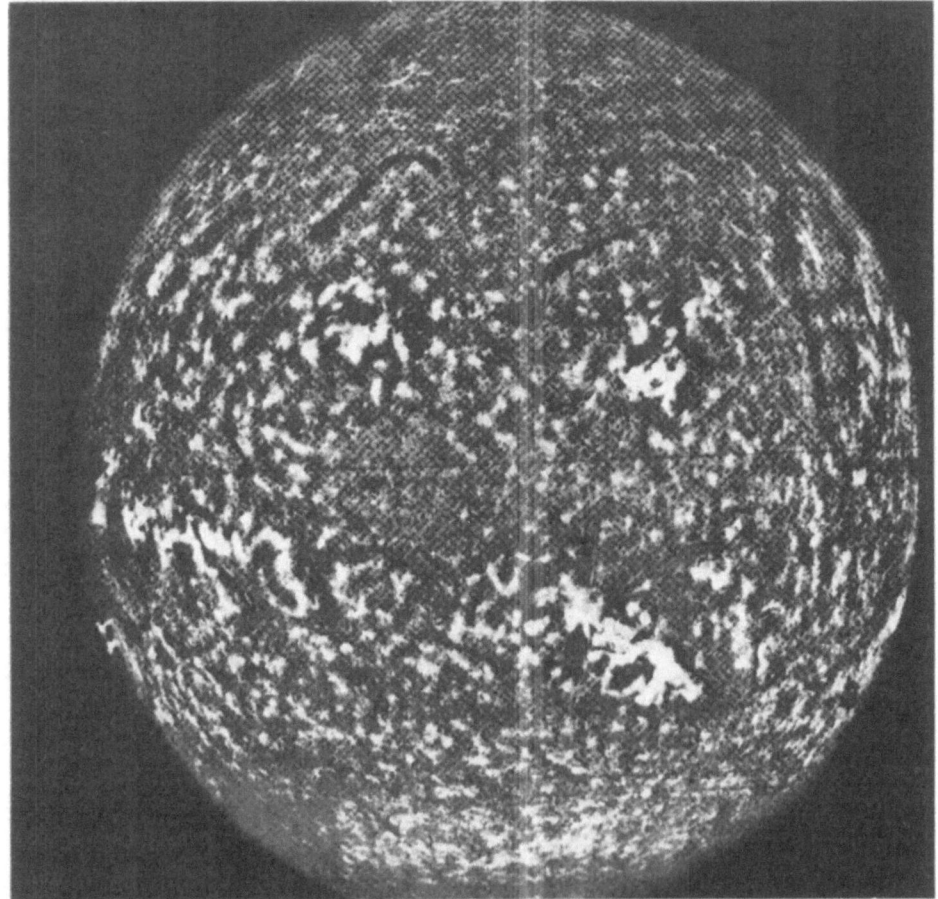

Fig. 44.   Photograph of the Sun in H_α light (6542 Å). (*Sacramento Peak Observatory photo.*)

A. SOLAR FLARES

Observation of the Sun in white light reveals only the presence of sunspots, which give information concerning its overall activity. By working in monochromatic light centered on the brightest emission lines, one can distinguish many surface inhomogeneities on photographs similar to that in Figure 44 (taken in the red light of H_α). The granular structure of the photosphere appears, with its brilliant *plages* (faculae) and dark filaments, which sometimes attain a length of hundreds of thousands of kilometers and an altitude of tens of thousands of kilometers. These are various types of *prominences*, for when they are seen at the edge of the disk, they stand out against the dark sky with various, more or less stable, shapes.

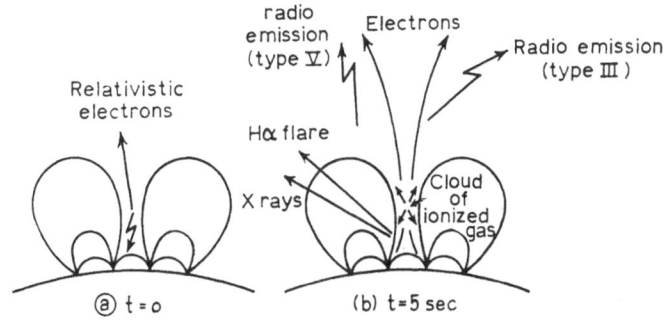

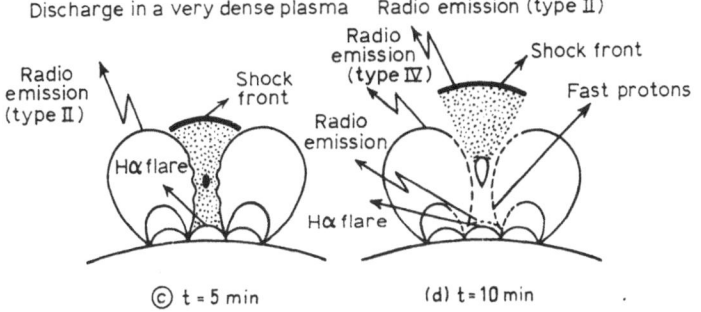

Fig. 45. Development of a solar flare according to J. P. Wild (1962).

All these solar events represent sources of abnormal emission of photons and of charged particles.

Sometimes large increases in luminosity are seen at certain points. This phenomenon is known as a solar flare.

A flare begins with an explosive phase in which the very intense emission, standing out from the continuum, extends rapidly over a large area. The source is always located in the vicinity of a sunspot.

The lifetime of a flare increases with the area it covers and with the linewidth of the enhanced H_α emission. In exceptional cases, one-thousandth of the solar disk can be covered by the flare. It then lasts for two or three hours, with a slow decline (Figure 45).

The energy released is considerable. It is estimated to be 10^{25} J during the first few minutes of an intense flare, or the equivalent of tens of millions of the most powerful thermonuclear bombs. This energy is not all in radiative form; some of it serves to accelerate the particles emitted, enabling a large number of them to escape from the attraction of the Sun.

B. IONOSPHERIC DISTURBANCES

A. *Causes*

The existence of the flare generally results in an enhancement of the D region. This is an effect with a sudden commencement, like the flare itself; it is abbreviated as SID (sudden ionospheric disturbance).

In a few minutes, one can observe a considerable increase in the electron density in the lower ionosphere, which is maintained for 15 min to 1 or 2 hrs. This effect involves the whole of the sunlit hemisphere, which indicates that it is the result of photon emission.

Although the optical observation of flares shows an excellent correlation with SIDs, it must still be noted that some flares are not followed by any disturbances at all; this is usually attributed to observational difficulties, for there is no *a priori* reason why a fraction of the photons released should not affect the Earth.

We already know that the upper atmosphere can be ionized only by short-wave-length photons ($\lambda < 1300$ Å). The enhancement of the 'visible' H_α line is therefore not enough to produce ionization. For a long time it was believed that the Lyman α line (1215 Å) behaved like H_α, and was sufficient to explain the phenomenon. However, this line would have to penetrate the atmosphere to an altitude of 60 km, which would require an unlikely increase in intensity and which, moreover, has been proved to be false by rocket observations.

It is now assumed that the 'SID' is due to an increase in X-ray radiation. Rockets have found greatly increased X-ray fluxes and a hardening of the spectrum, leaving hardly any doubt as to the correctness of this interpretation.

B. *Effects*

The 'SID' can be observed by its effects on the propagation of radio waves. These effects vary with frequency and therefore with the altitude at which they take place, and they yield information concerning the state of the disturbed ionosphere. They concern waves ranging in frequency from several kilohertz to about a hundred megahertz.

As waves of low and very low frequency are transmitted by a 'metallic' reflection from the D region, a sudden phase anomaly is observed at the time of the SID, revealing a rapid lowering of the level of this reflection by as much as 15 km. Moreover, the signal is enhanced at reception just like the signals from lightning discharges, which include low and very low frequencies.

In the megahertz region, on the other hand, a strong fade-out is observed, with the absorption increasing by a factor of 5 to 10. The same thing happens to the cosmic radio noise.

At very high frequencies (from 30 to 100 MHz) the connection between transmitter and receiver is possible only through scattering from the ionization irregularities of the lower atmosphere. We naturally observe differences at these frequencies also, either enhancement or fade-out. This depends in practice on the penetration of the radiation from the solar flare, and thus on the intensity of the SID. If the SID is weak, the electron density is increased only down to the level of the scattering. An enhancement results. If the SID is strong, the increase in electron density goes down to lower levels where collisions are more numerous. Then there is fade-out.

The SID is often marked on magnetograms by crochets of small amplitude (20 to 30 γ). These magnetic crochets exist only for rather intense flares which take place near the solar meridian facing the Earth. They are observed over an entire hemisphere and do not depend on the position of the Sun. They are attributed to ionospheric currents, which do not necessarily circulate at the same altitude as those that produce the regular variations in the terrestrial magnetic field.

2. Corpuscular Emission from the Active Sun

All other types of disturbances are due to the emission from the Sun of charged particles, which are superimposed on the usual solar wind. Two proofs of this fact have been known for a long time:

– these disturbances produce terrestrial effects which depend strongly on the geomagnetic latitude;

– these effects appear only after a transit time of two or three days, which is not at all comparable to the 8 min required by photons to reach our planet.

As we have seen, the solar corona is completely ionized and extends well beyond the orbit of the Earth. This ionized region moves outward from the Sun and is constantly renewed. The principal constituents are electrons and protons, traveling at the same velocity. The density of the corona has been measured out to about 20 solar radii by taking advantage of total eclipses of the Sun. Beyond this distance it is poorly known, except in the vicinity of the Earth, where it has been measured by space probes. The density is on the order of $10^9/cm^3$ at the base of the corona, and decreases regularly to as little as 10 or $20/cm^3$ at the outskirts of the terrestrial magnetosphere.

At the time of a solar eruption, *a storm plasma*, both denser and more rapid, is added to the normal solar plasma. It is difficult to give a figure for the resulting increase in density. A factor of 10 to 100 seems reasonable.

Two techniques are used to study eruptions on the Sun: optical and radio. The eruptions take place over a center of activity whose development in its different aspects is indicated by sunspots, flares, and prominences. An examination of the prominences indicates that matter rises and falls back, guided by what there is every reason to suppose are lines of force of local magnetic fields. Part of this matter is brought to high altitudes from which it can escape.

The study of solar radio noise confirms this particle emission and gives information concerning the mechanism by which it takes place. Although the Sun constantly emits electromagnetic signals in the radio region, it has been known for 25 years that flares are accompanies by 'bursts' at meter and decameter wavelengths. Taking into account the ionization profile of the corona and what we know about the plasma frequency, it is readily seen that to any distance h from the Sun there corresponds a cutoff frequency $f(h)$, as in the various regions of the ionosphere. This frequency decreases rapidly with increasing distance from the Sun. A wave emitted at frequency f will be received at the Earth only if it is emitted at a distance from the Sun greater than the value h defined by $f=f_c(h)$.

Plotting the intensity of the emission as a function of frequency and time, we obtain a graph like Figure 46, on which we can recognize different types of signals arriving with different delays. The delays can be connected with the velocities of plasmas ejected from the Sun. The type III burst corresponds to relativistic plasma oscillations

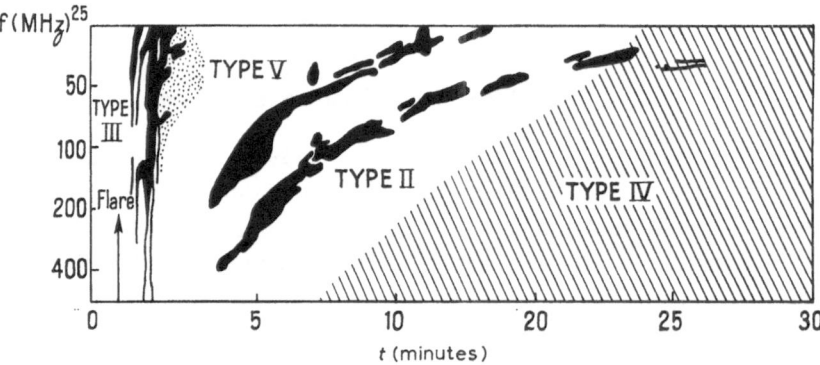

Fig. 46.   Histogram of a solar burst (J. P. Wild, 1963).

and electron velocities. They are much too rapid to be correlated with the appearance of magnetic storms. The type II burst was long associated with storms. But it implies unacceptably low velocities for the storm plasma, and appears much less often than terrestrial disturbances. It seems that the type IV burst, which is of long duration (sometimes as much as one day), is directly connected with the progress of the storm plasma through the solar corona. It is always associated with flares.

In order to understand the motion of this plasma in space, we must return to Parker's solar wind. This wind has a kinetic energy density greater than the magnetic energy density of the interplanetary field. The field is pushed back by the plasma, which itself carries along the solar magnetic field from the region in which it arose, the lines of force remaining anchored to the Sun. Since the emission is not purely radial, on account of the solar rotation, the lines of force have the form of Archimedes' spirals which intersect the orbit of the Earth at an angle somewhat different from $\pi/2$ (Figure 47).

This theoretical picture is confirmed by the observations, particularly those carried out by *IMP 1* outside the magnetosphere and the recent ones of *Pioneer VI*, which was put into orbit around the Sun in December 1965. Measurements of the field in magnitude and direction reveal a structure in sectors. Those which are marked with plus signs in Figure 47 indicate a field moving away from the Sun. This structure

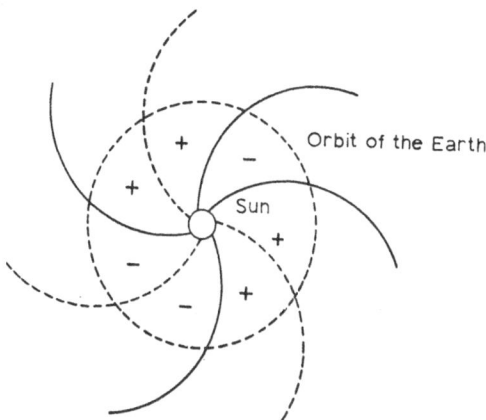

Fig. 47. Lines of force of the solar magnetic field dragged along
by the solar wind (J. M. Wilcox).

rotates with the Sun in twenty-seven days. The changes of direction along the spirals have not yet been explained.

What happens to these lines of force at great distances from the Sun? Logically they should close upon themselves at a distance which cannot be precisely stated. Moreover, distortions appear during the measurements, and it is reasonable to suppose that they increase beyond the orbit of the Earth.

This is the field and plasma distribution encountered by the storm plasma, which draws an additional field from the active regions of the Sun. Since it has a velocity greater than that of the solar wind, it pushes before it the normal plasma and field and advances in the form of a tongue, according to the model of Gold (Figure 48).

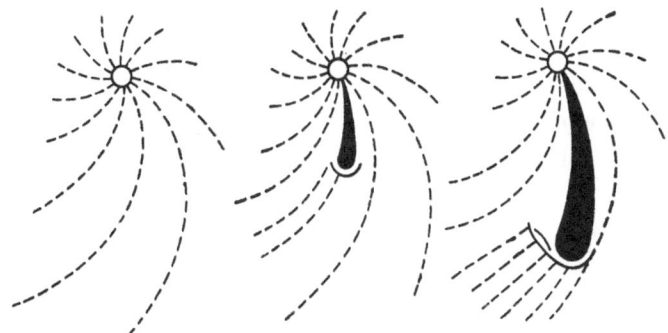

Fig. 48. Progress of the storm plasma (after T. Gold).

This flow is of the supersonic variety, whence we deduce the existence of a shock front moving ahead of the plasma which may account for the sudden commencement shown by many storms on Earth.

Our planet does not necessarily encounter the storm plasma. This depends on the lateral extent of the plasma and on the location of the source. If a direct encounter does not take place, the magnitude of the resulting disturbances depends upon the size of the solar flare and upon the way in which the Earth approaches the plasma.

The proofs of this schematic theory are still not very numerous, for few space probes have had an opportunity to study these effects. However, several spacecraft directed towards the Moon and towards Venus have encountered large temporary fluxes moving away from the Sun before the development of a storm with a sudden commencement on our planet. Thus Mariner II was able to detect a storm at a distance of 8 500 000 km from the Earth five hours before it reached us; its plasma density was greater by a factor of 6 than that ordinarily observed, and the energy of its charged particles was greater by 25%. We also recall the effect of the solar wind on comet tails, mentioned in Chapter V. We would expect the orientation of the comet tails to be changed during the solar events described above, and that is in fact what we observe. Finally, we note the *Forbush effect*, which concerns interstellar cosmic rays. These rays consist of very energetic charged particles, sometimes attaining energies of 10^{14} MeV. Most of them are protons and helium nuclei, which must pass through the solar corona before reaching us. We would expect their flux to be modified by the state of the corona, with an excess of solar activity producing a decrease in flux. Just such a decrease is observed, slowly in the course of the solar cycle and abruptly at the time of solar flares. It was long believed that this effect was due to a distortion of the terrestrial magnetic field; but since the same effect has been observed by spacecraft beyond the magnetosphere, it has been concluded that this modification is not of terrestrial origin.

3. Solar Cosmic Rays and Polar Cap Absorption

The interstellar cosmic rays are not to be confused with the solar cosmic rays, which have been identified ever since 1942 as protons emitted by the Sun. It is only quite recently, in 1961, that it has been possible as a result of the work of the International Geophysical Year to establish a clear correlation between these cosmic rays and the interruptions of radio communications above the polar regions which we shall call polar cap absorption.

The two effects are governed by the same solar event. They concern only high latitudes.

A. SOLAR COSMIC RAYS

Our knowledge of this domain comes mostly from measurements made by balloon, with ionization chambers, Geiger counters, and nuclear emulsion plates.

These instruments have identified protons and helium nuclei, but have not been able to observe high-energy electrons. The fluxes are isotropic and the energy spectra are of the form $KE^{-\gamma}$, where γ is on the order of 4. Most of the particles are high-energy protons (at least 80 to 100 MeV).

Since balloons can go no higher than 30 km of altitude, measurements by rockets and satellites were necessary. They made it possible to detect at higher altitudes protons of lower energy, going down to 1 MeV. We recall that the discovery of the Van Allen belts was the result of a project to study the solar protons. Since then, satellites placed in polar orbits have studied the phenomenon systematically; the *Discoverers*, for example, had become very radioactive at the time of their recovery as a result of this effect.

When we follow the intensity of this radiation over a long period, we observe three types of variation:

– a slow variation with a period of eleven years, which indicates a maximum number of events around the minimum of the solar cycle;

– a decrease in intensity during several consecutive days coinciding with magnetic storms. This is the Forbush effect, which we have already mentioned;

– an abrupt amplification by a factor of 1000 or 10 000 at the time of very intense solar flares. This very rare phenomenon occurred in particular on February 23, 1956, with exceptional strength. One hour after the flare, it was followed by a very strong ionospheric absorption over a vast geographic area, at geomagnetic latitudes of greater than 60°, having a very large diurnal variation.

These observations lead us to think that we are dealing with the arrival of a large number of charged particles from the Sun, producing a large ionization excess in the lower ionosphere. The role of the terrestrial magnetic field is obvious. It deflects the protons towards higher latitudes.

This type of absorption is very distinct from the one we encountered with the SID, particularly in its duration and its localization. It also differs from the absorption produced by magnetic storms, as we shall see in the following section.

Before we can understand the ionospheric effects, we must consider the behavior of such a proton flux in the presence of the atmosphere and of the terrestrial magnetic dipole. We can apply the Störmer theory, which neglects interactions. Recalling what we have already said on this subject, we see that the radius of gyration of the proton depends on its energy. The expression for this radius is

$$R_g = \frac{mv}{ZeB},$$

where $Z = 1$ for the proton; particles having the same charge and the same momentum behave in the same fashion.

The charge can be eliminated by introducing the concept of *rigidity*

$$\rho = \frac{W}{Z},$$

where W is the kinetic energy calculated with the relativistic effect taken into account, for the cosmic rays have large velocities upon arrival.

Störmer showed that to each magnetic latitude λ there corresponds a critical rigidity

$$\rho_c = 14.9 \cos^2 \lambda,$$

where ρ_c is expressed in BeV and is such that if $\rho < \rho_c$, the particle cannot reach the

Earth at vertical incidence. When ρ is a little less than ρ_c, the particle can still reach the Earth at a certain angle, but this phenomenon is negligible in the first approximation.

On the other hand, passing through the entire atmosphere requires an incident energy greater than 2 BeV – that is, a rigidity of about 3 BeV. Such protons occur only at latitudes greater than 50° (Figure 49). They cause ionization at the expense of their kinetic energy, leaving behind them a trail of ions and electrons according to a well-known law of nuclear physics. The ions created are much more numerous towards the end of the trajectory.

Since the incident charged particles have a broad energy spectrum, many of them

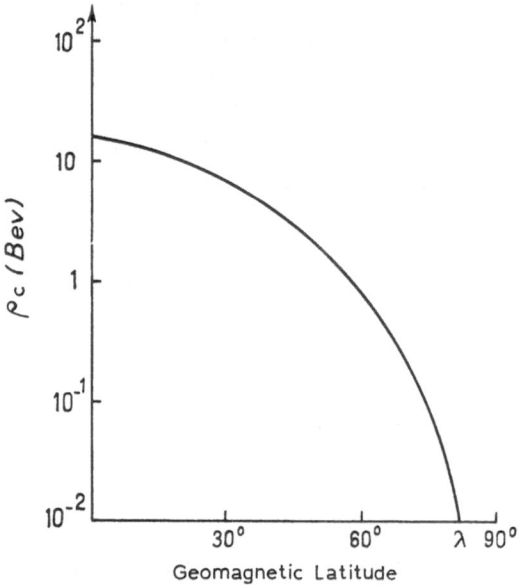

Fig. 49.   Critical rigidity for protons in the terrestrial magnetic field (cos² λ law).

are halted before reaching the ground, leaving a considerable electron density in the atmosphere, especially at low altitudes. These distributions can be calculated taking into account recombinations, captures, etc.; and the process of photo-dissociation caused by the presence of the Sun produces a large diurnal variation. One can predict a maximum number of electrons at around 75 km during the day, and at around 85 km at night.

B. POLAR CAP ABSORPTION IN THE IONOSPHERE

The measurement of the absorption of waves in the ionosphere is a complicated problem. For a long time only vertical sounders were used, making it possible to obtain the minimum frequency giving rise to an echo and the amplitudes of the successive reflections at a given frequency.

This method has several disadvantages:

– the results depend upon the performance of the equipment, which differs from one station to another and varies in time;

– the 'dynamic range' is limited. In the case of strong absorption, no more echoes are obtained, for the minimum frequency goes below the critical frequency of the F region.

Nevertheless, many results have been obtained by this method. They allow us to make statistical studies, but are less interesting when it comes to particular events.

A new technique appeared with *riometers*. The transmitters are natural extra-terrestrial sources. Their frequency must be greater than the cutoff of the F region, so that this method complements the preceding one. The principle consists of continuously comparing the signal received with a local diode noise. The difference between the signals is used for an automatic adjustment of the two levels. After calibration, the current of the diode gives the absorption.

Finally, we recall the use of propagation at high frequencies by irregularities in the ionosphere due to turbulence and to meteor trails. If the locations of emission and reception coincide, we are working by back-scattering. If this is not the case, the

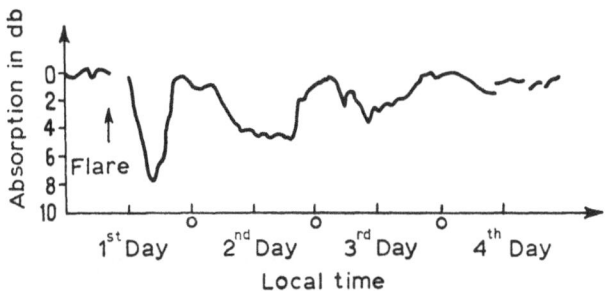

Fig. 50.   Polar cap absorption in Canada. Reception frequency = 30 MHz.

signal is received by forward scattering. If the irregularities responsible for the scattering are located above a source of significant absorption, this type of propagation yields data concerning the absorption.

A typical abnormal absorption in the polar regions takes place with a delay of 3 to 4 hours after the original solar flare. This time is too long for rectilinear propagation of the protons in space, thus confirming Parker's theory that the lines of force are curved. Moreover, the delay depends on the position of the flare on the Sun.

The radio phenomenon begins abruptly, with the absorption increasing for several hours. The effect continues for several days, but it is strongly modified by the diurnal variation (Figure 50). The absorption is smaller at night, as is easily understood when we consider the origin of the D region. The variations recorded at sunrise and at sunset are particularly interesting, for they enable us to study the effect of visible photons as a function of the altitude reached by the rays of the Sun before sunrise and after sunset. The absorption has been observed to involve levels as low as about 40 km, which appears to be in conflict with the electron distributions calculated and cited above.

We may note that at sunrise the tangential rays of the Sun have to traverse a considerable thickness of the atmosphere, but this explanation is not sufficient. Another screening effect must be found. It may be provided by the presence of the ozonosphere, for this layer cannot be traversed by any photon of wavelength less than 3000 Å. Such photons will therefore not be available to cause photodissociation. This explanation requires the presence at low altitudes of a dominant molecular ion which has not been identified. It cannot be O_2^-, for this ion is sensitive to photons that can traverse the ozonosphere.

The long duration of polar cap absorption events presents other problems. The direct effects of a flare last at most two hours. But the absorption effects detected sometimes extend over more than a week. We must therefore assume either that the protons are emitted by the Sun over an equally long time or that there is a storage region somewhere near the Earth.

The first alternative is rather unlikely, for it is the flare that supplies the energy. Moreover, retaining this hypothesis would also require us to assume that in some cases the Earth receives protons emitted from the far side of the Sun.

The second interpretation is hard to imagine in the polar regions, where there are no belts. Injection and release mechanisms would also have to be found.

Current thinking is directed towards the existence of a magnetic barrier in the solar system which would be capable of reflecting protons towards the Earth, thus greatly increasing their travel time. This barrier, which would have to be located well beyond our planet, remains to be discovered.

4. Magnetic and Ionospheric Storms

A. MAGNETIC ACTIVITY

The terrestrial magnetic field can be resolved into three components – the horizontal and vertical components and the declination. More than 100 magnetic observatories constantly record the values of these components. The annual means show a slow variation called the secular variation, but there are other variations of much shorter period. Diurnal, seasonal and lunar components can be observed; these may originate in the upper atmosphere, if we assume the existence of a system of electric currents governed by the Sun and to a lesser degree by the Moon.

Besides these systematic variations, other variations are observed which tend to affect the entire surface of the Earth and which have been given the name of *magnetic storms*. We have seen that they originate in individual solar events.

Magnetic activity, giving the degree of such disturbances, is indicated by a planet-wide index which results from a comparison of the recordings of all the observatories. The Greenwich day is divided into intervals of 3 hours, each one having its own index K_p. To obtain the index, one measures the amplitude of variation of the most active component of the magnetic field, and attaches to it a value of K between 0 and 9. As there is a strong latitude effect, the absolute scale is not the same for each observatory, and the assignment of 'K' requires experience and judgment.

It is clear that these values of K depend upon the local time, which must be eliminated to arrive at a worldwide index. A dozen stations situated at magnetic latitudes between 48° and 63° are used to establish standard indices K_s. The mean of the K_s yields K_p.

A magnetic storm is observed as a high value of K_p, although it is not possible to fix a clear threshold, for the storm may be more or less marked. Above all, it is defined by a characteristic development of the components of the field.

Magnetic storms are accompanied by *ionospheric storms*, which are more difficult to describe. Since the F region is the site of many anomalies, as we have seen, its maximum electron density and ionization distribution do not constitute sufficient criteria. It is therefore unrealistic to try to average them over time and place in order to arrive at a planet-wide index comparable to K_p. Storms are very complex phenomena which result not only in modifications of the field, but also in modifications of the temperature of the neutral and charged particles, and in optical and X-ray emission in certain regions. Thus at high latitudes a storm manifests itself in a spectacular fashion in the polar aurorae, which we shall describe in the next section.

There are two principal types of storms: *recurrent* storms and *sporadic* storms.

The statistics of storms indicate periodicities of 25 to 30 days, which can correspond only to the rotation of the Sun. This rotation is not uniform but depends upon latitude, so that there is a certain indeterminacy in the period – which is a function of the position of the centers of activity that create the storm. Storms are, as we have said, caused by the emission of additional solar plasma, which can reach the Earth more or less directly. Some of these eruptions last for weeks or months and thus throughout several successive solar rotation periods. The recurrence ceases either because the source of the emission disappears or because the Earth emerges from the plasma beam. The sources of recurrent storms are not always identified. In some cases they are flares and prominences, but the emission can continue after they have disappeared for a terrestrial observer.

When a comparison is made with the Wolf number, a good correlation is found. However, we must note that storms continue to occur during the years of the solar minimum, and that the maximum magnetic activity occurs two years after the maximum solar activity.

It is not always easy to associate a sporadic storm with a center of solar activity, and more particularly to establish a one-to-one correspondence between flares and storms. It takes several days for the storm plasma to reach us, and during that time many other flares can occur. However, it has been observed that the most intense flares produce the most intense storms, especially when they are located near the solar meridian facing us.

B. DEVELOPMENT OF A MAGNETIC STORM

Let us follow the progress on Earth of a storm, with its different phases. This study can be individual or statistical, for storms often show a comparable development more or less extended in time and more or less intense, which makes it possible to group them into categories.

Geophysicists have made an effort to classify storms; the usual practice is to distinguish between storms with a sudden commencement and storms with a slow commencement. When the commencement is sudden, 'rising fronts' from 1 to 10 min are recorded at several stations. Nevertheless, there are sometimes different impulses; and these can also take place without being followed by a storm, or during a storm. Almost all strong storms are of this type. It was once thought that they were necessarily connected with flares, but this is very uncertain, so that their distribution remains debatable. More interesting is the discovery of the mixed storm, or storm associated with the arrival of solar cosmic rays. The radiation has practically no effect on the magnetic field, but changes the D region considerably, making the interpretation of the associated ionospheric storm much more complicated.

The description of a storm requires the introduction of two space coordinates – magnetic latitude and longitude (λ and φ) – and time, counted from the beginning of the storm. The longitude is measured in the normal fashion from the magnetic meridian

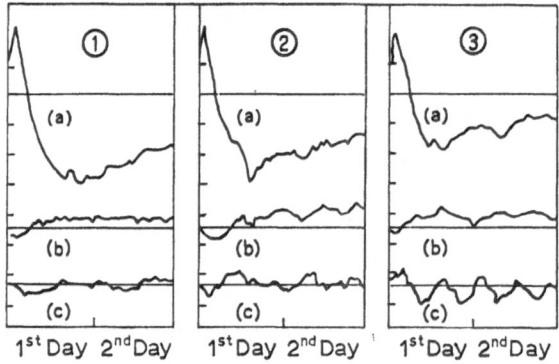

Fig. 51. A magnetic disturbance at three latitudes (S. Chapman). (a) Horizontal component; (b) Vertical component; (c) West declination. (1) Low latitudes; (2) $\lambda = 40°$; (3) $\lambda = 53°$.

of Greenwich, but it is more meaningful to refer it to the midnight meridian which turns with the Sun. In latitude one can distinguish three regions in each hemisphere:
- the low-latitude region;
- the auroral region;
- the polar region.

Any component of the field measured on magnetograms can be divided into two parts. The first is the mean value provided by all the stations at the same magnetic latitude, and the second is a residual which reveals a longitude inequality. We shall be interested mostly in the first part.

Figure 51 depicts the development of the horizontal (H) and vertical (V) components and the declination (δ) at different latitudes. The horizontal component has by far the most pronounced variations. It changes in the same way no matter what the latitude. Studying the development of this component alone, we can detect several phases:
- the beginning of the storm, marked by an abrupt rise;
- a so-called initial phase, in which H remains above its normal value for several hours;

– a main phase, which corresponds to a strong depression in H, with the minimum being reached at the end of a day;

– a final phase, which brings H very slowly back to its original value.

In general, these phases are observed to be shorter, the more severe the storm. This is shown in Figure 52.

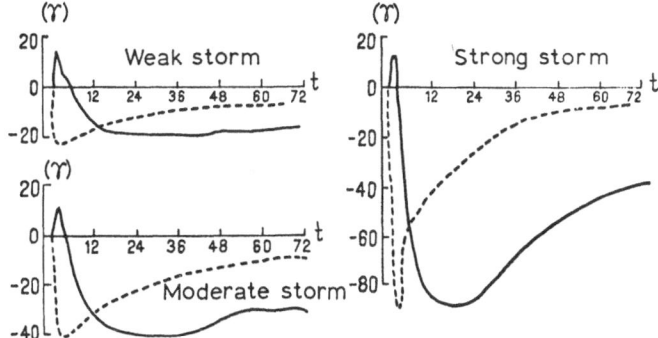

Fig. 52.   Variation of the horizontal component during storms of different intensities.
————— mean component; – – – – – residual component (cf. p. 90).

C. THE IONOSPHERIC STORM

It is convenient to keep in mind the different phases of the magnetic storm while reviewing the ionospheric effects observed by the usual methods (vertical sounders from the ground and from satellites, riometers, oblique propagation, etc.).

We observe first of all an abrupt enhancement of the D region, which produces sudden ionospheric absorptions comparable to those described in connection with the solar cosmic rays. These absorptions are present mainly in the auroral zones. The F region, on the other hand, is scarcely modified during this phase, at least insofar as we can judge from soundings, which are far from continuous.

The initial phase is marked by a disturbance of the F region, which seems to set in smoothly. It is difficult to compare the change in the F region with what would take place in the absence of a storm. This region varies too much from day to day and from place to place for us to determine a reference behavior. Moreover, the disturbance causes the appearance of diffuse signals, spread out in altitude, which make any reading of the frequency or of the altitude of reflection difficult, if not impossible. We do not yet know, for example, if the change in sign of the variation of the horizontal component causes changes in the ionosphere.

During the main phase, the ionospheric regions are profoundly modified.

The degree of ionization of the E region increases, and the disturbances in the F region depend strongly on latitude.

At high latitudes, there is depopulation, while at tropical latitudes, the contrary is observed. This is true not only at the level of maximum ionization, but also for the total ionization content; we are therefore dealing with a real phenomenon, and not merely a redistribution of the charged particles. No satisfactory explanation exists as yet. The temperature increase caused during the storm by the precipitation of energetic particles, brought about by the arrival of hydrodynamic waves, changes the rate of production of electrons, the loss mechanisms, and the group motions of the charged particles. This heating has been proved to be real, for an increased braking of satellites which orbit at the corresponding altitudes has been observed.

The explosions of thermonuclear devices in 1962 produced controlled disturbances, in the course of which the critical frequencies increased along the trajectory of a pressure wave, moving outward from the explosion center with a velocity of 400 m/s.

5. Polar Aurorae

The polar aurora is a phenomenon of natural night-sky luminescence, which can generally be observed only at high magnetic latitudes. We are familiar with it from descriptions and photographs.

Aurorae can take on varied forms and colors. An effort has been made to classify them, but the nomenclature suffers from a certain degree of empiricism.

We shall limit ourselves to distinguishing the three principal forms: quiet arcs, rayed arcs, and diffuse patches.

Arcs are thin luminous sheets inclined at the same angle as the terrestrial magnetic field, extending over one hundred to several hundred kilometers in altitude, and sometimes over several thousand kilometers in the horizontal direction. The rayed arcs are striated along the lines of force, and sometimes move rapidly to form curtains.

In the last ten years or so, observational techniques have been improved and extended. In addition to the studies of visible aurorae carried out by cinematography and spectroscopy, radio aurorae have also been studied; they do not necessarily coincide with the optical aurorae. Radio soundings make it possible to obtain echoes just as from an ionospheric layer.

An aurora can last several minutes or a whole night. Drawing on the surface of the Earth curves of equal probability for the appearance of an aurora in the course of a year, we obtain ovals centered on the geomagnetic poles. The line along which this probability is greatest is located at 23° from the pole in the northern hemisphere and at 18° in the southern hemisphere.

Figure 53 indicates the position of this line in the northern hemisphere. We see that the Canadian stations are particularly well placed.

An aurora begins with the appearance of one or more arcs perpendicular to the direction of the pole. The arcs closest to the pole may rise higher in the sky to come closer to it, while those which are farther away descend. This expansion in latitude causes the aurora to cover a wide region. Next there appear irregularities which move along the arcs, principally in the eastward direction. Towards midnight the aurora loses its organized structure.

The horizontal extent of the arcs was not known until the International Geophysical Year in 1958, when the perfection of the 'all-sky' camera made it possible to obtain photographs covering all directions at once. A series of these cameras was placed from coast to coast in Canada, and showed that certain aurorae were spread out over 5000 km.

An aurora borealis is always accompanied by an aurora australis. The association is not always obvious, for the magnetically conjugate regions are not always precisely known.

Solar disturbances are the cause of the great majority of the polar aurorae. The above description makes it clear that the terrestrial magnetic field plays an important role, so that we would expect the aurorae to originate in charged particles. When the accompanying storms are intense, aurorae develop down to magnetic latitudes of about 20°. There is good correlation with sunspots, leaving no doubt that an aurora is due to the interaction of charged particles with the upper atmosphere. From the

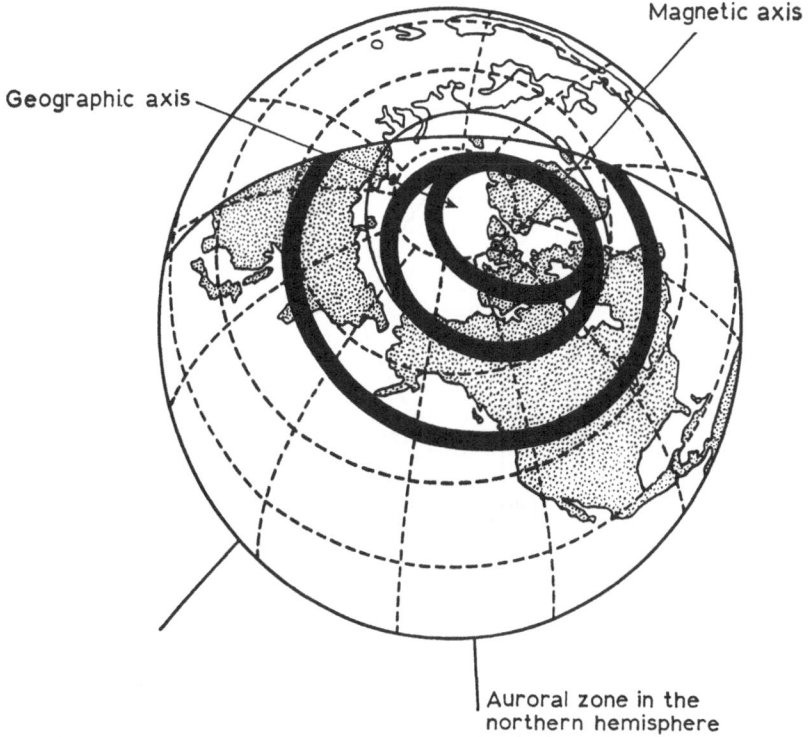

Fig. 53. Line of maximum frequency of aurorae at three phases of the solar cycle
(after S. I. Akasofu).

statistics, one can recover the eleven-year cycle in the frequency of appearance and in the latitudes involved. The twenty-seven-day periodicity can also be observed.

Most aurorae appear green or blue-green with pink or red borders. Identification of the spectral lines and bands shows that the emission comes from atomic oxygen (5577 and 6300 Å) and from molecular nitrogen (3914–4700–6500 and 6800 Å), which is not surprising.

However, a study of the altitude distribution shows that the two oxygen lines do not come from the same regions (Figure 54). This difference can be explained by the inequality of the lifetimes of the excited states of oxygen. The state from which the green line arises has a lifetime of 0.7 s; the other one has a lifetime of 200 s. At an altitude of about 100 km, the air is too dense for the latter state to be de-excited before the intervention of many collisions.

A spectroscopic examination indicates that the base of aurorae is located at the level of the E region of the ionosphere, with slight variations depending on the strength of the phenomenon. The upper limit is less well defined. Some aurorae reach altitudes of 1000 km.

There are two other types of aurorae (A and B) which have not yet been mentioned, for they rarely occur. The former occurs only occasionally in the course of a solar cycle. It is characterized by a diffuse red light at an altitude of several hundred kilometers, and accompanies a violent magnetic storm connected with a strong ionospheric absorption in the polar regions. It has been attributed to the arrival of high-energy solar protons in the atmosphere. Type B is characterized by the presence

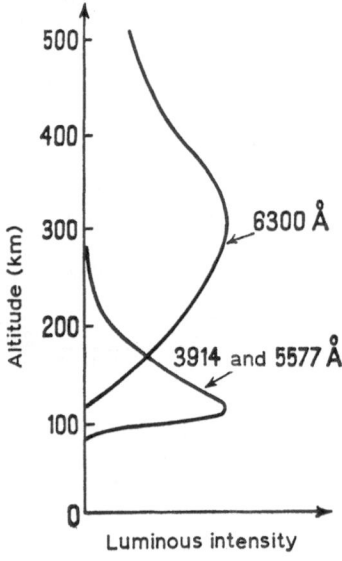

Fig. 54.   Auroral emission as a function of altitude.

of a purple fringe at the base of an active aurora. It represents a very rare phenomenon at low altitudes.

The hydrogen lines have been observed in certain auroral spectra. When they are observed in the direction of the magnetic zenith, they are broadened, asymmetric, and displaced towards the ultraviolet. The asymmetry disappears along the magnetic horizon. At the time of this discovery, it was believed that protons were responsible for aurorae; but the hydrogen lines are not always present. We must conclude that protons accompany certain aurorae but are not the principal cause of the auroral arcs.

Direct studies at high altitudes by means of balloons, rockets, and satellites have proved in the last few years that the luminosity of the aurorae is due primarily to electrons with energies of several kilo-electron volts; the proton flux is much less important.

While sounding balloons cannot reach the altitude where the emission takes place, they have been able to observe the electrons indirectly. Although the electrons go no lower than 80 km, they emit X-rays which can be recorded by detectors at much lower altitudes. Many measurements, both north and south of the auroral belts, have proved that this X-ray emission is well correlated with the presence of rayed aurorae. Within the auroral belts, the X-ray emission is constant and much less intense. If we estimate the total number of electrons necessary to create a large aurora, we find that it is ten times greater than the number contained in the Van Allen belts. This tends to prove that although the aurora empties the belts, the accompanying storm more than replenishes them.

6. Theory of Storms and Aurorae

A. STATEMENT OF THE PROBLEM

For a long time specialists have been seeking a satisfactory theory of storms and aurorae. Many ideas and calculations have been published, but the conclusions have had to be rejected one after the other. They do not account for all the phenomena observed at ground level, in the ionosphere, and in more distant regions of space; and they often contain internal contradictions.

Once we have recognized that storms and aurorae are the consequences of the arrival of a supplementary plasma in the vicinity of the Earth, we must still explain:

– how this plasma can penetrate to the interior of the magnetosphere, which in theory is well protected;

– how organized systems of worldwide currents – whose induction, combined with that already existing, accounts for the observed variations – can be quickly established;

– how the electrons can descend as far as the lower ionosphere, where they cause the ionizations and recombinations observed during the aurorae and the ionospheric disturbances which accompany them.

Aurorae have a longer history than storms, for they could not help attracting the attention of observers. Many great names of the past are associated with them: Hall, Dalton, Cavendish, Ångström, etc.

Gauss was the first to speak of an electrical phenomenon, but the geophysical knowledge of his time was not sufficient for any progress to be made. The first serious theories appeared at the end of the last century, with Birkeland and Störmer.

The study of magnetic storms also began with the work of these two physicists. Since the space between the Earth and the Sun was naturally assumed to be a vacuum, Störmer studied, as we have seen, the behavior of a relatively small number of charged particles emitted from the Sun over a short period and interacting with the terrestrial dipole. The resulting model is unique and can really be applied only to solar cosmic rays. Next, Chapman and Ferraro examined in detail the interaction between a plasma jet and the terrestrial magnetic field, and thus explained the initial phase of storms. From a detailed analysis of the variations of the horizontal component of the terrestrial magnetic field, Chapman derived the concept of the value averaged over longitude

(DSt), which is not sensitive to the diurnal variation, and of the residual value (DS), which provides more information.

More recently, Dessler and Parker rediscussed the interaction in hydromagnetic terms. We shall confine ourselves here to the recent model of W. I. Axford and C. O. Hines, which was formulated in 1961 and has been further developed since then. It has the merit of attempting to provide a single satisfactory theory to explain all the geophysical anomalies of the upper atmosphere.

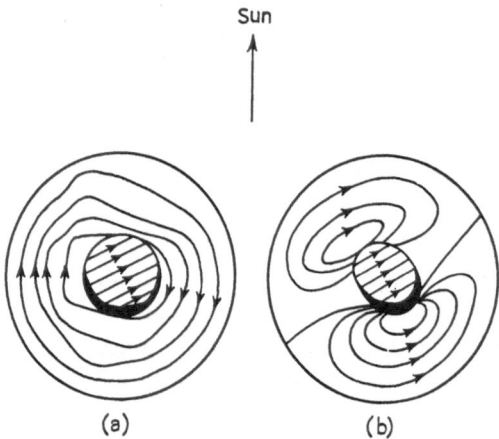

Fig. 55.   Systems of currents which explain the variation of the terrestrial magnetic field during a storm.

(a) Currents during the main phase. (b) Mean currents during the magnetic bays (after Silsbee and Vestine).

B. THE THEORY OF AXFORD AND HINES

The authors begin by noting that there is no reason to suppose that the magnetosphere is static, even in the absence of disturbances. In the case of the ionosphere, we have already seen that we must consider the motion of the charged particles if we wish to account for the behavior of the F region. Why should systematic motions not exist at much greater altitudes?

Axford and Hines assume that there is large-scale convection inside the magnetosphere, arising from the interaction of the solar wind with the terrestrial magnetic field. They assume that the coupling is viscous, although this hypothesis is not essential for their conclusions.

Referring to Figure 41, we observe that the lines of force of the terrestrial magnetic field can be classified into two groups:

– those at low latitudes. They close upon themselves and do not depart greatly from the lines of force of a dipole;

– those at high latitudes, which extend outward to form the tail of the magnetosphere.

The lines of force which fall between these two categories on the sunlit side are greatly deformed, and it is difficult to place them in one or the other of the two groups.

At ground level they correspond to a 'zone of confusion' in each hemisphere, symmetric with respect to the magnetic equator.

Any increase in the solar wind transfers the lines of force from one category to the other, so that the 'zones of confusion' approach the magnetic equator.

As a consequence of the plasma-induction interaction on the magnetopause, the solar wind carries the charged particles belonging to the edges of the magnetosphere towards the geomagnetic tail. These charged particles, which are trapped in the terrestrial magnetic field, accumulate behind the Earth. Such a state of affairs cannot

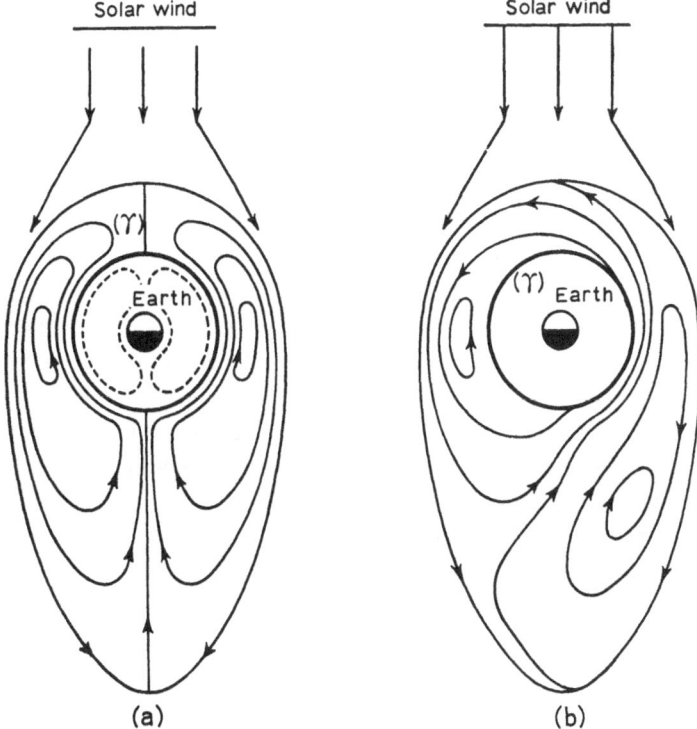

Fig. 56.   Magnetospheric convection in the equatorial plane. (a) The rotation of the Earth is not taken into account. (b) Distortion of the lines of flow when the rotation of the Earth is taken into account (after W. I. Axford and E. O. Hines, 1961).

continue unless the charged particles return towards the Earth through the interior of the magnetosphere, as is indicated in Figure 56. This convection must take place along the lines of force, to which the charged particles are 'frozen'. The particles located at lower altitudes are carried along by a convective current of the same type.

The phenomenon can be presented in a different but equivalent manner by stating that the solar wind sets the charged particles of the outer magnetosphere in motion by establishing a velocity field V. The motion takes place in the terrestrial magnetic field in such a way that the Laplace force $(V \wedge B)$ is balanced by the electric field E due to the accumulation of charged particles, which is equivalent to writing that:

$$E + V \wedge B = 0.$$

Since the electric field can be derived from a constant potential φ, we also have $\mathbf{E} = -\mathbf{grad}\ \varphi$.

We deduce that the velocity $\mathbf{V}$ and the induction $\mathbf{B}$ (normal to $\mathbf{E}$) are tangent to the equipotential surfaces. Thus the magnetic lines of force are electric equipotential lines.

However, there are effects which tend to oppose the motions of the magnetosphere, especially near the Earth; for atmospheric viscosity and ohmic losses become appreciable. The velocity of flow is fixed by the balance between the energy furnished by the solar wind and the energy dissipated in this manner.

The observations lead us to situate the inner border of the magnetospheric convection at a distance of four or five terrestrial radii, corresponding to lines of force which intersect the ground at latitudes near those of the auroral zones. It may therefore be assumed that the convection does not concern lower latitudes – where, in fact, the observed disturbances are much less marked.

This border is indicated in Figure 56a, drawn by Axford and Hines for the case of a closed geomagnetic cavity. In addition, this cavity is assumed to be stable because of the existence of the outer Van Allen belt.

The solid lines in Figure 56a indicate the convection in the equatorial plane, while the dashed lines indicate an 'internal' circulation which is needed to explain the system of currents associated with the (DS) component of which we have spoken.

In this figure the Earth is assumed to be motionless. In reality, the ionosphere and the charged particles of the magnetosphere are carried along by the Earth, causing the distortion indicated in Figure 56b.

If we take into account the gyration of the charged particles and the curvature of the lines of force, we find new drift motions which are significant for the most energetic particles (electrons above several keV). Since these longitude drifts are in opposite directions for protons and electrons, while the convection does not depend on the sign of the charge, charged particles of the same energy are separated and do not circulate at the same level around the Earth. This decoupling is especially apparent at the time of disturbances.

One last phenomenon remains to be mentioned. Experiments made by satellites have shown that the outer part of the magnetosphere is turbulent. This turbulence, carried along by the convection, will affect lower altitudes without being damped, because the convection causes a compression of the plasma. Moreover, this compression brings about an acceleration of the particles which can produce severe disturbances in the case of enhancement of the solar wind.

Figure 56, which presents equatorial cross-sections, shows that the convection does not involve ionospheric altitudes at low latitudes, for the 'protection' circle (γ) has a radius of several terrestrial radii. This is no longer the case when one approaches the polar regions, for the lines of force which are tangent to (γ) approach the ground and reach it at auroral latitudes. At the level of the ionosphere, one must take into account collisions between charged and neutral particles which stop the ions (and, to a lesser extent, the electrons) from moving on the equipotential surfaces ($\varphi = $ const) defined above. There is, in fact, a force of viscous friction which depends upon a collision frequency k_1 and the motion of the neutral particles.

In the absence of the latter factor, the change takes place when the ion gyrofrequency becomes less than k_1, which is the case in the lower part of the E region. The velocity of the ions is less than that of the electrons, which become the principal constituent of the current. In reality, there *is* motion of the neutral particles; but contrary to classical dynamo theory, they are set in motion by the ions, provided there is sufficient time – several hours in the E region, but much less in the F region.

C. GEOPHYSICAL CONSEQUENCES

Axford and Hines have shown how the general ideas in the preceding paragraphs can provide a basis for the explanation of many geophysical disturbances, when the effects of turbulence, convection, and compression are examined in turn.

The effects of the magnetospheric turbulence should, as we have explained, be reflected at low altitudes in magnetic and ionospheric observations of magnetic activity, auroral activity, increased ionospheric absorption, and diffuse echoes from the ionosphere. All these phenomena have in fact been observed in a certain range of geomagnetic latitudes, and especially at night.

Convection produces a compression of the plasma which causes an acceleration of the charged particles.

The solar wind contains relatively mono-energetic charged particles, but turbulence quickly broadens the spectrum from a fraction of an electron volt to several kiloelectron volts. Convection acts upon all the particles, but there is a difference between low-energy particles and high-energy particles – the former follow the previously described equipotential surfaces, while the others leave them more or less quickly because of their much more marked drift motions.

In the first case, the compression is adiabatic. If we ignore losses by precipitation in the lower atmosphere, we see that when the plasma is transported from a geocentric distance r_1 to a geocentric distance $r_2 < r_1$, the volume of the plasma undergoes a change $(r_1/r_2)^3$, which must be multiplied once more by the factor r_1/r_2 to take into account the shortening of the lines of force guiding the motion of the plasma. There results an energy increase of

$$\left(\frac{r_1}{r_2}\right)^{3\,(\gamma_1 - 1)} \left(\frac{r_1}{r_2}\right)^{(\gamma_2 - 1)}$$

where the ratio γ, connected with the degrees of freedom, is not necessarily the same in the two factors. It can be shown in this way that energy gains of 10 or more are possible.

In the second case, the energy increase is less, for the shortening of the lines of force is limited by the drift motion. The maximum energy attained depends on the level. It is on the order of 10 keV at auroral altitudes.

Whatever the details may be, convection acts as a constant source of particles on the night side of the magnetosphere, and the electrons accelerated in this way can reach the altitudes of the aurorae. Their precipitation depends upon the position of their

mirror points (cf. Chapter II), which are lowered during convection according to the
relation $\sin^2\theta/B = C_1$, as already given. In particular, we have at the mirror point

$$B_m = \frac{B_{eq}}{\sin^2\theta_{eq}} = B_{eq}\left(\frac{v^2}{v_\perp^2}\right)_{eq},$$

where B_{eq} is the induction above the equator. B_m increases with the energy of the
charged particle, while the line of force on which the particle moves is displaced
towards lower altitudes. There is thus a considerable lowering of the mirror points.
Since electrons drift towards the east and protons towards the west (Figure 9), we
would expect to observe a precipitation of protons in the evening and electrons in the
morning, which is just what experiments have shown.

Magnetic storms produce temporary variations in the components of the terrestrial
magnetic field. We have seen how the horizontal component, which is the most
significant, changes with time. Its variation can be analyzed into two parts at a given
latitude, during the main phase:

– the part which does not depend upon the longitude (the mean value DSt);
– the residual part (DS) which depends strongly on the longitude.

The first of these components depends so little on latitude that its origin must be
located far from the ground, in the form of an annular current circulating westwards
in the magnetic equatorial plane, at an altitude of several terrestrial radii. Störmer
had already postulated the presence of this current in connection with the aurorae.
Modern theories retain the hypothesis of its existence. But it must be admitted that
space probes have not yet observed it in an unambiguous fashion.

The second component appears to be due to currents circulating in the atmosphere,
particularly above the auroral zones, eastwards in the evening and westwards after
midnight. It is tempting to think that this system is produced by the arrival of a plasma
storm and is superimposed upon the system of electric currents which circulates in the
atmosphere and explains the regular variations (Chapter III).

One of the first points to be cleared up concerns the existence of the sudden commencement. Certain storms are originally marked, as we have seen, by an abrupt
change in the field which sometimes takes place in one minute, although it takes a
whole day for the plasma to arrive from the Sun. This phenomenon leads us to assume
that the storm plasma has a sharp leading front, which hits the magnetopause and
pushes it back. The impact can give rise to hydromagnetic waves which carry the
magnetic disturbance along the lines of force at the Alfvèn velocity, on the order of
100 km/s. A detailed study of the distribution in latitude and longitude of the precise
moments of the sudden commencements would serve to give a firmer basis for this
explanation. Other experiments prove that at the instant of the sudden commencement, electrically charged particles strike the atmosphere in large numbers, so that
hydromagnetic waves are not the only phenomenon involved.

It is assumed that the initial phase is a direct consequence of the arrival of charged
particles at the magnetopause, as Chapman and Ferraro suggested. Since we have
learned that the region occupied by the terrestrial magnetic field is filled with plasma,
we have had to introduce a propagation time before the effect can be detected at the

ground. Propagation in the magnetosphere must account for the fact that the variation is much more marked at higher latitudes. As for the main phase, it is explained by an increased circulation of trapped particles in the form of the equatorial current already cited, at least for the equatorial latitudes, and by the auroral current system for other latitudes. All these enhancements, however, are poorly understood.

Axford and Hines have shown that the compression of the magnetospheric ionization, leading to an increase in the particle energy, can bring about such enhancements; for new charged particles are captured in large numbers by means of convection in the magnetosphere, which feeds the current systems. In this perspective, the few hours that separate the beginning of the storm from the beginning of the main phase can be considered as the convection time.

CHAPTER VII

MICROMETEORITES

1. Micrometeorites and Their Detection

Up until now, we have spoken of neutral or ionized gases, the gravitational field, the magnetic field, and solar radiation. We have still to consider the fragments of solid matter which do not belong to the Earth, but which it continually encounters in its journey around the Sun. We are speaking, as the reader may have guessed, of meteorites.

The terminology varies somewhat among different authors. We shall restrict the term *meteors* to the luminous traces produced by meteorites upon contact with the atmosphere, and we shall call objects which reach the ground *meteorites. Micrometeorites* are in principle too small to survive the journey through the atmosphere. The smallest of them form the *meteoritic dust.*

On a clear night, the naked eye of an experienced observer can count about ten 'shooting stars', which are seen as transitory luminous trails. These trails are produced by solid bodies, basically silicates and ferromagnetic metals, which penetrate the atmosphere with velocities of several tens of kilometers per second on contact with the air.

As we have said, only the largest of them partially survive atmospheric erosion, and arrive at the ground after having undergone evaporation and often fragmentation. Observations have made it possible to estimate the frequency of fall as one meteorite per year per million square kilometers. We recall that the surface area of the Earth is 500 million square kilometers.

Large meteorites are extremely rare cases. It is only every 100 000 years that the Earth receives meteorites of 50 million kg, capable of leaving vast craters like that in Arizona.

For a long time, meteorites were detectable only through their interaction with the atmosphere. Pushing back the visibility threshold with optical instruments, one can count them in much greater quantities. The use of very sensitive telescopes and of photography has become common. Observations are now made with so-called super-Schmidt cameras, which have both a large field and a high degree of sensitivity. They are provided with a rapidly rotating shutter, so that the meteor trails are interrupted at a known frequency, making it possible to measure the velocity and the acceleration of the meteorite.

By observing the same trails simultaneously from several points on the Earth's surface, one can reconstruct the trajectories. Semi-empirical formulae make it possible to derive the mass from measurements of the luminous intensity. If I is the intensity and m the mass of the meteorite, we have

$$m = \frac{2\,It}{v^2\tau},$$

where t is the lifetime of the trail and τ is the fraction of energy converted into luminous energy.

We often speak of magnitude rather than luminosity, as for the other heavenly bodies. The magnitude M is defined by the relation

$$M - M_0 = -2.5 \log_{10}\left(\frac{I}{I_0}\right),$$

which converts a ratio into a difference. An arbitrary zero of apparent magnitude has been chosen, essentially corresponding to the luminosity of the star Vega. A magnitude difference of -1 corresponds to a logarithmic decrease of $(100)^{1/5} = 2.512$. Very luminous objects have negative magnitudes. That of the Sun is -26.7; that of the full Moon is -12. Optically, the observer can detect meteors down to magnitude 3 or 4.

The Earth is believed to capture 110 kg of matter per magnitude per day, considering only objects of diameter greater than 100 μ.

Just as the polar aurorae are observable by radio as well as optical methods, meteorites can also be detected by radar. For the trails are actually fine columns of plasma about ten kilometers long and containing 10^{10} to 10^{12} electrons/cm. These columns diffuse rapidly, and permit transitory reflections of radio waves. This is a powerful method of studying the trails, and it has been developed considerably over the last twenty years. One works at a fixed frequency between 30 and 100 MHz – much higher than the highest critical frequency for the ionosphere – either with a continuous beam or with 500 to 1000 pulses per second.

When the ionized column is dense enough, the trail can be considered to be a veritable metallic cylinder. This is the case with visible meteors and even fainter ones, down to magnitude $+6$. The reflected signal then comes from several Fresnel zones located at the nearest point to the emitting beam directed at the trail. Operating from several bases at once, one can deduce the altitude, the velocity, and the distance of the meteor. The radar method reaches to magnitude $+10$, and works as well by day as by night, in all kinds of weather. This method also gives the atmospheric density and temperature at the altitudes in question.

All the meteors that can be analyzed from the ground appear at altitudes between 70 and 170 km, with a maximum at the level of the E region of the ionosphere. One might wonder whether they contribute significantly to the ionization of this region. The question has been much debated. It is certain that by day the contribution is negligible. By night the prevailing electron density is on the order of 10^3 electrons/cm³. It is conceivable that meteors play a certain role in the appearance of the so-called sporadic E layer* at times of especially numerous impacts, and that they participate in the maintenance of the normal region.

In the last few years, artificial satellites have offered new possibilities in this domain also. They make direct observations possible at altitudes where the terrestrial atmosphere no longer has any effect. By means of satellites, the extent of the observations has been considerably broadened to take in the meteoritic dust formed by the micrometeorites that surround the Earth. We can now identify, count, and analyze fragments of matter of extremely small mass, down to 10^{-13} g: if the optical terminology were still meaningful, these would have magnitudes of $+36$. For such small masses, the solar radiation pressure plays an important role.

It has been necessary to develop new methods of analysis, which actually go back 18 years to the time of the first geophysical rockets. Several types of apparatus have been perfected:

– microphonic sensors. A piezo-electric crystal is mounted on a wall whose outer surface receives the meteoritic bombardment. Each impact gives rise to a voltage signal which is amplified and analyzed. Impacts outside the sensitive surface must not be recorded;

– conducting wires buried in an insulating support, and covering a certain portion of the surface of the spacecraft. The wires are attached to a voltage supply and mounted in parallel. A meteorite of sufficient mass leaves a large enough hole in the insulator to break a wire, producing a signal;

– pressure-sensitive collectors. Sealed pockets are filled with gas under a certain pressure. When they are perforated, a signal results;

– photomultipliers, which measure the luminous intensity emitted during the impact of a microparticle with sufficient velocity. Fragments of the size of a micron have thus been measured. From the intensity and the duration of the flash, the kinetic energy can be deduced. In practice, the impact takes place on a lucite or glass surface placed in front of the photocathode and protected from the surrounding light by a thin deposit of aluminum. This procedure makes it possible to go down to 10^{-13} g;

– microcondensers. When the sheathing is penetrated, the material is vaporized, producing a conducting gas that discharges the condenser. The condenser is recharged in a few milliseconds, giving rise to an electrical signal.

All these instruments must be carefully calibrated. This is done in the laboratory by dropping balls of well-determined mass and radius onto different points on the sensitive surface. The drop height is variable. For low velocities, glass balls with diameters between 50 and 500 μ are used; they fall from heights varying between several millimeters and several centimeters. Their masses go from a few micrograms to 200 μg. The impact velocities vary from 20 to 400 cm/s. The collisions are therefore elastic. One must either correct for the effect of air resistance or create a vacuum.

* We have not mentioned the existence of this layer during our description of the ionosphere. In contrast to the other layers, it is present at a given place only in an aleatory fashion, as its name indicates. Located at around a hundred kilometers of altitude – and thus in the heart of the normal E layer – it has a small thickness and does not always completely screen the higher layers. Many studies of this phenomenon have been made. None of the interpretations proposed for its origin and its behavior can be entirely accepted as yet.

These velocities are very far from those to be expected with micrometeorites. Explosion-powered pistons impart velocities of up to 4 km/s to steel balls with diameters between 30 and 300 μ. The collisions become inelastic, and leave craters from which the energy and momentum transfer can be deduced. To obtain higher velocities, up to 100 km/s, one must have recourse to other methods, such as macroparticle accelerators.

2. Meteoritic Astronomy

Meteorites obey the law of universal gravitation. If the force of radiation pressure (which is relevant only for microparticles) is neglected, their motions satisfy Kepler's three laws. In order to determine their origin, we must know the center of attraction of their orbit. We must also know whether the orbit is closed (an ellipse) or open (a hyperbola). It is now practically certain that all meteorites belong to the solar system.

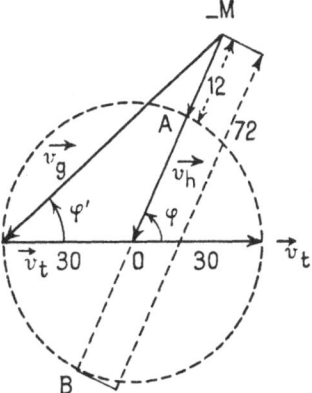

Fig. 57. Analysis of the geocentric and heliocentric velocities of a meteorite.
$|\mathbf{v}_g|$ is no greater than MB = 72 km/s and no less than MA = 12 km/s.

From Newton's law it can be shown that the velocity of a mass m orbiting around another, much greater mass M, satisfies the equation

$$v^2 = \frac{2K}{r} + C_1,$$

where K is the product of the mass M and the constant of universal gravitation: 1.3×10^{11} km³/s² for the Sun and 4.10×10^3 km³/s² for the Earth, the two centers of attraction which are possible *a priori*. The distance between M and m is r.

C_1 is a constant determined from the reconstructed orbit. If $C_1 > 0$, the orbit is hyperbolic.

At the Earth's distance from the Sun, the velocity v cannot be greater than 42 km/s if the object belongs to the solar system. And in fact, it has been observed that all the particles have a velocity less than this critical value.

The velocities measured by a terrestrial observer are relative to the Earth, or geocentric velocities. There is a simple vector relation between this velocity $\mathbf{v}_g$ and the heliocentric velocity $\mathbf{v}_h$. We have (Figure 57)

$$\mathbf{v}_h = \mathbf{v}_g + \mathbf{v}_t,$$

where $\mathbf{v}_t$ is the velocity of the Earth in its orbit, or 29.8 km/s. The figure shows that

$|\mathbf{v}_g|$ lies between 12 and 72 km/s. The smallest velocity (12 km/s) is still greater than the escape velocity from the Earth. Thus all meteorites coming from outside follow, in principle, hyperbolic orbits relative to the Earth. Some of them may, however, be slowed down sufficiently to go into orbit around the Earth.

When measurements are made from a satellite, an additional correction must be made to take its geocentric velocity into account.

At each instant there is a point on the surface of the Earth called the *apex*, which is located in the direction of the Earth's motion around the Sun.

Assuming that there is *a priori* an equal probability of seeing meteorites appear at all points on the sky, a terrestrial observer will be able to detect half of them – that is, those which are above the horizon. In fact, he will see a few more than half because of the motion of the Earth. The additional fraction is largest when the observation takes place at the apex itself.

The apex is not a fixed point on the globe. It describes a circle in 24 hours because of the rotation of the Earth, and this circle moves between the tropics from one day to the next, as the seasons progress. Thus there should be a seasonal variation in the number of meteorites recorded at any given point.

Whatever the observer's location, his distance from the apex is a minimum at about 6 a.m. local time, and this is the time at which the number of meteorites detected is greatest. This daily variation is confirmed when optical and radar records are reduced and compared.

On the other hand, it is well known that at certain times of the year meteorites are more numerous. This is the result of the Earth's passage through zones in which the meteorite concentration is higher than normal. Meteorites are divided into families. All the members of a single family seem to come from a localized region on the sky called the *radiant*. When determined from the Earth, however, this radiant – the point of intersection of the meteorite stream and the celestial sphere – is not the true radiant, for the motion of the Earth must be taken into account. Figure 57 shows that the true radiant φ is related to the apparent radiant φ' by the formula

$$\varphi = \varphi' + \arcsin\left(\frac{v_t}{v_h}\sin\varphi'\right).$$

A family of meteorites bears the name of the constellation located near its radiant at the time of year at which its meteors appear. Thus we have the Perseids in August, the Leonids in November, etc.

A parent comet can be attributed to some of these groups. The meteorites descended from it originally follow parallel orbits around the Sun, but because of the gravitational perturbations of the major planets – especially Jupiter – there is a slow diffusion which distributes these meteorites along belts around the Sun. The groups whose appearances are quite regular come from very old comets. However, these 'recurrent' meteorites make up only a very small fraction of all the meteorites observed.

3. The Direct Study of Micrometeorites

The study of micrometeorites began with the launching of the first American satellite, *Explorer I*. It carried a microphonic system. *Sputnik III* made a similar study, and since that time many spacecraft have been equipped for the detection of micrometeorites. From a list which has already grown very long, let us cite *Explorer VII, VIII, XIII, XVI* and *XXIII*; *Vanguard III*; the *Pioneers*; *Mariner II* and *IV*, etc. On the Soviet side, the *Luniks, Venusiks,* and some of the *Cosmos* series have also made it possible to study the cosmic dust in the neighborhood of the Earth, the Moon, and Venus.

Other things being equal, the greater the sensitive surface directed towards space, the more efficient the detection. This cross-section has continually grown. From a small

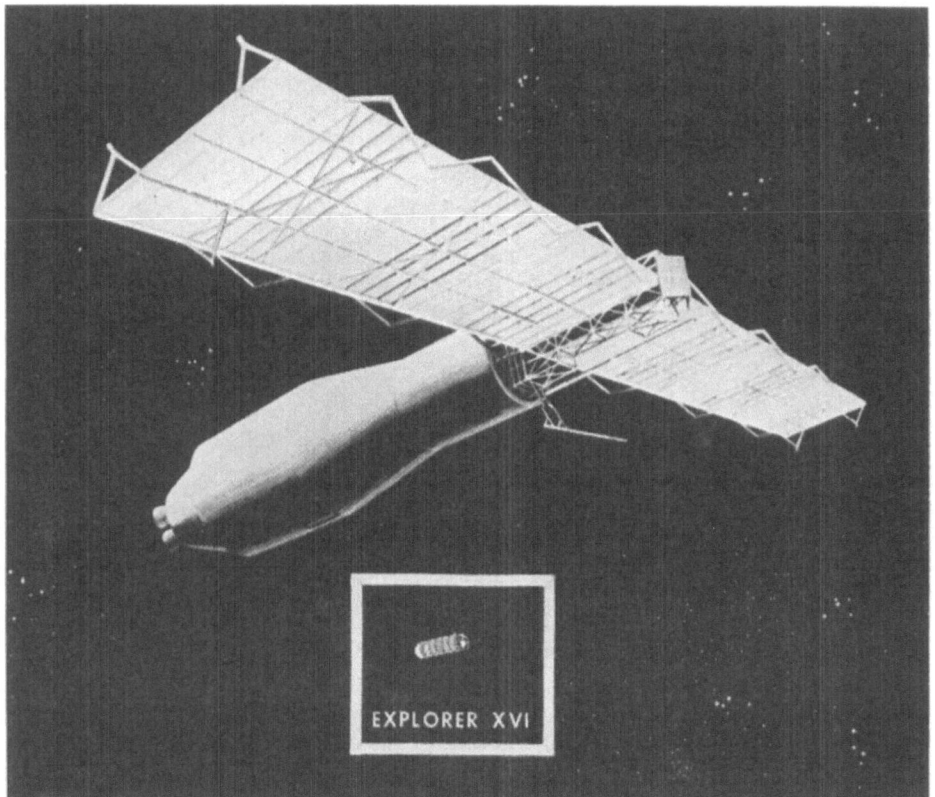

Fig. 58.   The *Pegasus* satellite, designed for the detection of micrometeorites.
(*NASA photo.*)

fraction of a square meter at the very beginning, it reached more than 250 m² with the *Pegasus* satellites (Figure 58). Three giant satellites weighing 1500 kg and called *Pegasus* were launched in 1965 with the rocket Saturn I, the first stage in the development of the giant rocket which was to enable the American astronauts to reach the Moon. Designed especially for the study of micrometeorites, they describe relatively low orbits: between 500 and 700 km for the first two, and 530 km for the third.

The *Pegasus* satellites carry two immense panels which are folded during launch and spread out after placement in orbit. These panels contain a large number of joined condensers on each of their faces. There are 416 condensers in all, divided into three categories depending on the thickness of the sheathing, so that the experiment is not limited to simply counting the microparticles. On the last of the *Pegasus* satellites certain elements are detachable, and a future astronaut is expected to bring them back so that a direct analysis of the impact traces can be made.

The number of perforations decreases as the thickness of the detectors increases.

The results obtained are important, for meteorites may be dangerous for long flights of large spacecraft, which must therefore be protected against them. The *Pegasus* satellites have given rather discordant figures, indicating spatial and temporal fluctuations in the meteorite density.

Using the measurements collected by all the satellites, it has been possible to draw a curve representing the cumulative flux per square meter per second as a function of velocity and mass. This flux is represented in Figure 59, where we use the magnitude

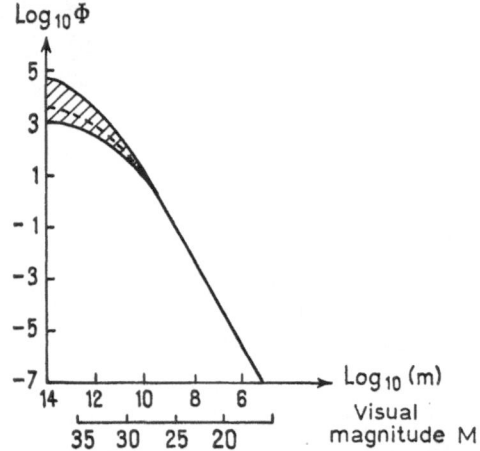

Fig. 59. Φ is the cumulative flux of meteorites per square meter per second, and m is the mass in grams.

scale (which below 10^{-2} g is useful for comparison purposes only). To go from magnitudes to masses is not a simple operation, for the calibration remains debatable. It is assumed that a meteorite with a mass of 1 g and a velocity of 30 km/s has a magnitude of 0. Measurements of very small masses with photomultiplier techniques reveal a large uncertainty – two orders of magnitude for 10^{-13} g.

The curve in Figure 59 is valid only in the neighborhood of the Earth. In the relatively dense layers of the atmosphere, measurements of particles whose size is about a micron may be false, for these particles may result from the fragmentation of larger meteorites. It is observed that the deceleration of faint meteorites increases much more rapidly than predictions which take into account only the atmospheric friction. Progressive mass loss is not sufficient to explain the phenomenon; the initial mass must be assumed to have a lower value than that deduced from the photographic observations. We must then suppose that meteorites break up or undergo surface erosion much more rapidly when they are of small size.

The slope of the curve thus obtained is not constant. It decreases in absolute value when the mass becomes less than a nanogram. This is due to the influence of solar radiation pressure, which represents an important force for very small masses. This force tends to sweep space free of these microparticles by pushing them out beyond the orbit of the Earth. The 'critical' mass, which can be defined in terms of the radiation pressure at the Earth's distance from the Sun, depends upon the density of the meteorite. This critical mass varies from 10^{-10} to 10^{-13} g. Micrometeorites of diameter less than 0 2 μ should be completely eliminated. If particles of this size remain present, they must be attributed to the debris of larger masses.

The results of Figure 59 give a mass distribution different from that obtained by extrapolating the results of terrestrial observations. The flux of micrometeorites

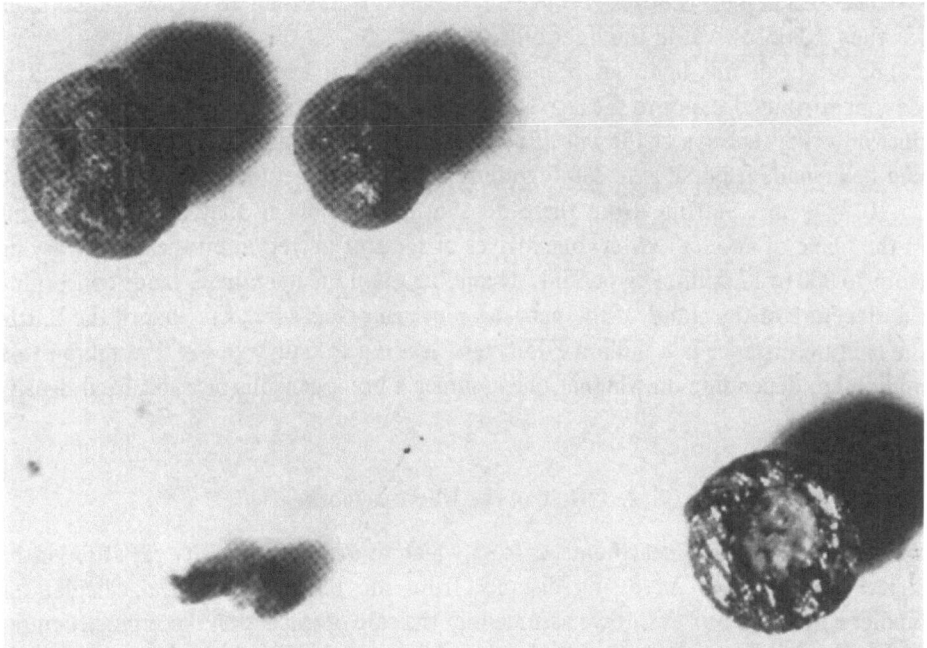

Fig. 60. Spheroids (magnified 300 times) recovered from ocean sediments. Columbia University. (*Scientific American photo.*)

becomes much larger, so that the mass of matter falling upon the Earth must have been underestimated in the past. This rate is apparently 10 million kg per day, with large variations. The dust surrounding the Earth must therefore be much more uniformly distributed than is indicated by the study of visible sporadic meteorites.

Research carried out by Petterson and Rotschi in the Pacific Ocean has shown that sediments corresponding to several million kilograms per day are deposited on the ocean floors, and that these sediments are of extra-terrestrial origin. It was an English oceanographer, J. Murray, who discovered as early as 1876 that the seabeds were covered with metallic spheroids. When examined under a microscope, these spheroids look like solidified spheres of melted metal (Figure 60). The distribution of these spheroids is now being studied as a function of region and depth. Soundings have been made to 15 m and deposits at the rate of 2 mm per thousand years have been observed, thus yielding the history of the phenomenon during 7.5 million years.

Other methods can be used to estimate the rate of accretion on our planet. Planes at high altitude collect dust, leading to higher figures than those given above. Probably a large part of what is collected in this way consists of particles lifted from the ground by the winds.

The observation of the zodiacal light is another indirect method of studying micro-meteorites. This light is observed after sunset in the tropics, towards the West, in the form of a pyramid-shaped glow whose luminosity decreases as the Sun goes farther below the horizon. As the Earth turns, this glow sinks. It disappears a few hours before sunrise, and then the same spectacle is observed in the East.

The zodiacal light is never visible when the Sun is less than 18° below the horizon, for then it is drowned in the light of the Sun.

The origin of this light has been much debated; as early as three centuries ago, Cassini attributed it, not to the terrestrial atmosphere, but to a cloud of interplanetary dust diffusing the rays of the Sun. Recent studies have confirmed this interpretation, and have made it possible to set a figure for the density between 5×10^{-21} g/cm^3 and 2×10^{-23} g/cm^3. Starting from these densities, and showing that this dust remains in the plane of Jupiter's orbit, one arrives at rates of accretion on the Earth varying from 70 000 to 17 million kg per day, depending upon the hypotheses made concerning the structure of the cloud. If the particles move in orbits similar to that of the Earth, the capture distance is 8 million kilometers, leading to daily rates of 1.5 million to 9 million kg, depending on whether one assumes a homogeneous or a localized density distribution.

4. Origin of the Micrometeorites

We still have very few measurements from which to derive the density variation of the meteoritic dust with increasing distance from the Earth. Theoretical calculations predict a decrease as $r^{-3/2}$. If we assume that the rate of impact on the superstructures of spacecraft is proportional to the density, at least for impacts of bodies heavier than a milligram, we obtain a density 100 000 times greater at an altitude of about 300 km than in the interplanetary cloud that gives rise to the zodiacal light.

It seems that there is a large concentration of meteoritic material near the Earth. The effect of the atmosphere certainly contributes to this, but not beyond a few hundred kilometers of altitude. Another explanation must be found. In addition to gravity, we must consider the solar radiation pressure and forces of electrical origin. Taking only gravity into account, one can explain the existence of a cloud around our planet with a maximum density at several thousand kilometers of altitude, forming a veritable belt which is constantly replenished from outside. Spacecraft have not yet observed it, but in the course of their journeys towards Venus and Mars, the *Mariners* proved that micrometeorites were 10 000 times more abundant in the neighborhood of the Earth than along their trajectories. On the other hand, the first results of *Luna X*, the Russian satellite put into orbit around the Moon, indicate a density 100 times greater near the Moon than in the space between the Earth and the Moon. According to certain authors, the ejection of material from the surface of the Moon when the craters were formed could account for the high density in the vicinity of the Earth. Meteorites leaving the Moon at velocities of 2 or 3 km/s, suitably oriented, could describe circumlunar orbits which would be transformed little by little into circumterrestrial orbits. It should be possible to verify this hypothesis in the near future.

Another mechanism to explain the concentration in the vicinity of a planet has been proposed. Collisions between particles, producing a loss of velocity, would place the meteorites in elliptical orbits; the particles themselves would come from comets and asteroids. Such collisions would have to take place in the Earth's sphere of influence, estimated at 200 terrestrial radii; for it is only within this sphere that terrestrial satellites can orbit about our planet. Beyond it, the perturbations of the Sun become dominant. The mass captured is thus proportional to the number of collisions, to the volume of the sphere of influence and to the probability of capture; for the kinetic energy after the collision must be smaller than the 'escape energy' at that point. This probability increases as one approaches the Earth, and the negative effect of the atmosphere can account for the measured densities. The hypothesis can be tested in the neighborhood of Venus, for example. This planet, like the Earth, is located within the zodiacal cloud; its gravitational acceleration is similar to that of the Earth, and it has no natural satellite.

BIBLIOGRAPHY

The present work represents only a preliminary exploration of the subject of the terrestrial environment. Space physics is a very young science, still in the course of development. The basic works written in the years previous to 1960 often seem outmoded in many fields, although they remain valuable. But there are not many recent works as yet.

As for original publications, they have grown spectacularly in number in the last few years. It is therefore out of the question to list them for readers who might wish to make a more detailed study of this vast subject. We shall limit ourselves to articles of a review nature. Certain opinions expressed in them may be put in doubt in the near future, as a result of space experiments now being planned or carried out. We shall refer primarily to publications in the *Reviews of Geophysics*, which make a particular effort to give an overall view. Articles by physicists in specialized fields are abundantly cited in these publications, and can easily be found through them.

GENERAL WORKS

(1) Mitra, S. K.: 1952, *The Upper Atmosphere*, 2nd ed., Calcutta, Asiatic Society.
(2) Ratcliffe, J. A. (ed.): 1960, *Physics of the Upper Atmosphere*, N.Y., Academic Press.
(3) Rawer, K.: 1957, *The Ionosphere*, N.Y., Frederick Ungar Publishing Co.
   These first three books deal principally with the ionosphere. The following books deal with much higher altitudes.
(4) De Witt, Hieblot, Lebeau (eds.): 1962, *Géophysique extérieure*, N.Y., École d'Été des Houches, Science Publishers.
(5) Alfvèn, H., Fälthammar, C. G.: 1963, *Cosmical Electrodynamics*, 2nd ed., London, Oxford University Press.
(6) Odishaw, H. O. (ed.): 1964, *Research in Geophysics*, Vol. 1: *Sun—Upper Atmosphere and Space*, M.I.T. Press.
(7) Hines, C. O., Paghis, I., Hartz, T. R., and Fejer, J. A. (eds.): 1965, *Physics of the Earth's Upper Atmosphere*, Prentice Hall Inc.

REVIEW ARTICLES

Chapter II

(1) A description of the Earth's magnetic field is the subject of Chapters III, VII, and VIII of a book by S. Chapman and J. Bartels, *Geomagnetism*, Oxford Press, 1962.
(2) The motion of charged particles in the magnetic field is treated in the following three publications:
   Northrop, T. G.: 'Adiabatic Charged-Particle Motion', *Rev. Geophys.* **1**, 283.
   Dragt, A. J.: 1965, 'Trapped Orbits in a Magnetic Dipole Field', *Rev. Geophys.* **3**, 255.
   Walt, M. and MacDonald, W. M.: 1964, 'The Influence of the Earth's Atmosphere on Geomagnetically Trapped Particles', *Rev. Geophys.* **2**, 543.
(3) For the behavior of electromagnetic waves in plasmas, the reader is referred to:
   Denisse, J.-F. and Delcroix, J.-L.: 1961, *Théorie des ondes dans les plasmas*, Monographie Dunod.
   Ratcliffe, J. A.: *The Magneto-ionic Theory*, London, Cambridge University Press.

Chapter III

As problems concerning the ionsphere are studied in detail in the first three general works, we shall merely cite:
   Friedman, H.: 1963, 'Ultra Violet and X-Rays from the Sun', *Ann. Rev. Astron. Astrophys.* **1**, 59, Palo Alto, California.

Evans, J. W. (ed.): 1963, *The Solar Corona*, Academic Press.

Reid, G. C.: 1964, 'Physical Processes in the D Region of the Ionosphere', *Rev. Geophys.* **2,** 311.

Herman, J. R.: 1966, 'Spread F and Ionospheric F-Region Irregularities', *Rev. Geophys.* **4,** 255.

Chapman, S.: 1956, 'The Electrical Conductivity of the Ionosphere', *Nuovo Cimento*, No. 4, Series 10, p. 1385;

and an article relating to the exploration of the upper side of the ionosphere by satellite:

Petrie, L. E.: 1963, 'Top-Side Spread Echoes', *Can. J. Phys.*, No. 41, p. 194.

Chapter IV

(1) On the Van Allen belts, the reader may consult:

Farley, T. A.: 1963, 'The Growth of our Knowledge of the Earth's Outer Radiation Belt', *Rev. Geophys.* **1,** 3.

Skuridin, G. A. and Pletnev, V. D.: 1965, 'Principal Hypothesis Concerning the Origin of the Earth's Radiation Belts', *Soviet Phys. Uspekki* **8,** 224.

Haymes, R. C.: 1965, 'Terrestrial and Solar Neutrons', *Rev. Geophys.* **3,** 345.

(2) The propagation of 'whistlers' is studied in:

Carpenter, D. L. and Smith, R. L.: 1964, 'Whistler Measurements of Electron Density in the Magnetosphere', *Rev. Geophys.* **2,** 415.

(3) Atmospheric tides are the subject of:

Wilkes, M. V.: *Oscillations of the Earth's Atmosphere*, Cambridge University Press.

Fejer, J. A.: 1964, 'Atmospheric Tides and Associated Magnetic Effects', *Rev. Geophys.* **2,** 275.

(4) For changes in the orbits of satellites, see:

Cook, G. E., King Hele, D. J., and Walker, D. M. C.: 1960, 'The Contraction of Satellite Orbits under the Influence of Air Drag', *Proceedings Royal Society A*, No. 257, p. 224.

Chapter V

Problems concerning the borders of the environment are discussed in the following articles:

Beard, D. B.: 1966, 'The Solar Wind Geomagnetic Field Boundary', *Rev. Geophys.* **2,** 335.

Hess, W. N., Mead, G. D., Nakada, M. P.: 1965, 'Advances in Particles and Field Research in the Satellite Era', *Rev. Geophys.* **3,** 521.

MacDonald, G. J. F.: 1963, 'The Escape of Helium from the Earth's Atmosphere', *Rev. Geophys.* **1,** 305.

Chapter VI

The effects of storms are described in:

Geomagnetism, Oxford, Clarendon Press, Chapter XII, 1962.

and those of the polar aurorae in:

Polar Aurorae, Oxford, Clarendon Press, 1955.

Among the theories recently proposed, we cite only:

Axford, W. I. and Hines, C. O.: 1961, 'A Unifying Theory of High Latitude Geophysical Phenomena and Geomagnetic Storms', *Can. J. Phys.* **39,** 1433.

Chapter VII

Levin, B. Ya: 1965, 'The Origin of Meteorites', *Soviet Phys. Uspekki* **8,** 224.

ASTROPHYSICS AND SPACE SCIENCE LIBRARY

Edited by

J. E. Blamont, R. L. F. Boyd, L. Goldberg, C. de Jager, Z. Kopal, G. H. Ludwig, R. Lüst,
B. M. McCormac, H. E. Newell, L. I. Sedov, Z. Švestka, and W. de Graaff

p.t.o.

16. S. Fred Singer (ed.), *Manned Laboratories in Space. Second International Orbital Laboratory Symposium.* 1969, XIII+133 pp.

17. B. M. McCormac (ed.), *Particles and Fields in the Magnetosphere. Symposium Organized by the Summer Advanced Study Institute, held at the University of California, Santa Barbara, Calif., August 4–15, 1969.* 1970, XI+450 pp.

18. Jean-Claude Pecker, *Experimental Astronomy.* 1970, X+105 pp.

19. V. Manno and D. E. Page (eds.), *Intercorrelated Satellite Observations related to Solar Events. Proceedings of the Third ESLAB/ESRIN Symposium held in Noordwijk, The Netherlands, September 16–19, 1969.* 1970, XVI+627 pp.

20. L. Mansinha, D. E. Smylie and A. E. Beck, *Earthquake Displacement Fields and the Rotation of the Earth. A NATO Advanced Study Institute. Conference Organized by the Department of Geophysics, University of Western Ontario, London, Canada, 22 June–28 June, 1969.* 1970, XI+308 pp.

21. Jean-Claude Pecker, *Space Observatories.* 1970, XI+120 pp.

22. L. N. Mavridis (ed.), *Structure and Evolution of the Galaxy. Proceedings of the NATO Advanced Study Institute, held in Athens, September 8–19, 1969.* 1971, VII+312 pp.

23. A. Muller (ed.), *The Magellanic Clouds. A European Southern Observatory Presentation: Principal Prospects, Current Observations and Theoretical Approaches, and Prospects for Future Research. Based on the Symposium on the Magellanic Clouds held in Santiago, Chile, March 1969, on the Occasion of the Dedication of the European Southern Observatory.* 1971, XII+189 pp.

24. B. M. McCormac (ed.), *The Radiating Atmosphere. Proceedings of a Symposium Organized by the Summer Advanced Study Institute, held at Queen's University, Kingston, Ontario, August 3–14, 1970.* 1971, XI+455 pp.

25. G. Fiocco (ed.), *Mesospheric Models and Related Experiments. Proceedings of the 4th ESRIN-ESLAB Symposium, held at Frascati, Italy, July 6–10, 1970.* 1971, VIII+298 pp.

26. I. Atanasijević, *Selected Exercises in Galactic Astronomy.* 1971, XII+143 pp.

27. Constantin J. Macris (ed.), *Physics of the Solar Corona. Proceedings of the NATO Advanced Study Institute on Physics of the Solar Corona, held at Cavouri-Vouliagmeni, Athens, Greece, 6–17 September, 1970.* 1971, XII+345 pp.

SOLE DISTRIBUTORS FOR U.S.A. AND CANADA:

SPRINGER-VERLAG NEW YORK, INC., 155 Fifth Ave., New York, N.Y. 10011